"十三五"国家重点出版物
出版规划项目

Series on Advanced Electronic Packaging Technology and Key Materials

先进电子封装技术与关键材料丛书

汪正平（C.P. Wong） 刘胜（Sheng Liu） 朱文辉（Wenhui Zhu） 主编

TSV 3D RF Integration
HR-Si Interposer Technology

TSV三维射频集成
——高阻硅转接板技术

马盛林（Shenglin Ma） 金玉丰（Yufeng Jin） 著

· 北京 ·

内 容 简 介

三维射频集成应用是硅通孔（TSV）三维集成技术的重要应用发展方向。随着 5G 与毫米波应用的兴起，基于高阻硅 TSV 晶圆级封装的薄膜体声波谐振器（FBAR）器件、射频微电子机械系统（RF MEMS）开关器件等逐渐实现商业化应用，TSV 三维异质射频集成逐渐成为先进电子信息装备领域工程化应用的关键技术。

本书全面阐述面向三维射频异质集成应用的高阻硅 TSV 转接板技术，包括设计、工艺、电学特性评估与优化等研究，从 TSV、共面波导传输线（CPW）等基本单元结构入手，到集成无源元件（IPD）以及集成样机，探讨金属化对高频特性的影响规律；展示基于高阻硅 TSV 的集成电感、微带交指滤波器、天线等 IPD 元件；详细介绍了 2.5D 集成四通道 L 波段接收组件、5~10GHz 信道化变频接收机、集成微流道散热的 2~6GHz GaN 功率放大器模块等研究案例。本书也系统综述了高阻硅 TSV 三维射频集成技术的国内外最新研究进展，并做了详细的对比分析与归纳总结。本书兼顾深度的同时，力求从较为全面的视角，为本领域研究人员提供启发思路，以助力我国在 TSV 三维射频异质集成技术研究的发展进步。

本书可供微电子先进封装以及射频模组领域研究人员、工程技术人员参考，也可供相关专业高等院校研究生及高年级本科生学习参考。

图书在版编目（CIP）数据

TSV 三维射频集成：高阻硅转接板技术=TSV 3D RF Integration：HR-Si Interposer Technology：英文/马盛林，金玉丰著. 一北京：化学工业出版社，2021.6

（先进电子封装技术与关键材料丛书/汪正平，刘胜，朱文辉主编）

ISBN 978-7-122-39484-2

Ⅰ．①T… Ⅱ．①马…②金… Ⅲ．①集成电路工艺-英文 Ⅳ．①TN405

中国版本图书馆 CIP 数据核字（2021）第 132609 号

本书由化学工业出版社与 Elsevier 出版公司合作出版。版权由化学工业出版社所有。本版本仅限在中国内地(大陆)销售，不得销往中国台湾地区和中国香港、澳门特别行政区。

责任编辑：吴　刚　李玉峰　毛振威　　　　　装帧设计：关　飞
责任校对：王素芹

出版发行：化学工业出版社（北京市东城区青年湖南街 13 号　邮政编码 100011）
印　　装：北京虎彩文化传播有限公司
710mm×1000mm　1/16　印张 18¼　字数 502 千字　2021 年 12 月北京第 1 版第 1 次印刷

购书咨询：010-64518888　　　　　　　　　　售后服务：010-64518899
网　　址：http://www.cip.com.cn
凡购买本书，如有缺损质量问题，本社销售中心负责调换。

定　价：298.00 元　　　　　　　　　　　　　　版权所有　违者必究

Preface to the Series

The technical level and development scale of the integrated circuit (IC) industry is one of the important indicators to measure a country's industrial competitiveness and comprehensive national strength, and is the source of modern economic development. The application of IC has already become routine in various industries, such as military satellites, radar, civilian automotive electronics, smart equipment, and consumer electronics, etc. At present, the IC industry has formed three major industrial chains of design, manufacturing and packaging testing, which have become the indispensable pillar in the IC industry.

IC packaging is an indispensable process in the IC industry, which is the bridge from chip to device and device to system. It is a key fundamental manufacturing part of the IC industry and a competitive commanding height for the core device manufacturing of the IC industry.

With the rapid development of IC technology, higher and higher requirements for miniaturization, multi-function, high reliability and low cost of electronic products are put forward. Facing this situation, the electronic packaging materials and technologies are undergoing rapid development, promoting lots of advanced packaging materials. Advanced electronic packaging materials and technologies are the core of IC packaging.

In order to promote the development of China's advanced electronic packaging industry and meet the urgent needs of researchers ranged from teaching and scientific study to engineering developing in the field of electronic packaging, the editorial committee has invited famous specialists to write the *Series on Advanced Electronic Packaging Technology and Key Materials* in recent years (English version). The series includes: "*Advanced Polyimide Materials*" "*From LED to Solid State Lighting*" "*Freeform Optics for LED Packages and Applications*" "*Modeling, Analysis, Design and Tests for Electronics Packaging beyond Moore*" "*TSV 3D RF Integration*" etc.

This series of books systematically describes the advanced electronic packaging from three aspects: advanced packaging materials, advanced packaging technologies and advanced packaging simulation design methods. This series covers the most advanced packaging materials such as polyimide materials and packaging technologies such as freeform optical technology, TSV (through-silicon via technology) packaging, and advanced packaging simulation design methods such as multi-physics analysis and applications. In addition, this series also makes a planning outlook and forecast for the development trend of advanced electronic packaging.

This series of books is of great worth for workers engaged in scientific research, production and application in electronic packaging and related industries, and also has great reference significance for teachers and students of related majors in higher education institutions.

We believe that the publication of this series of books will play a positive role in promoting the development of China's IC industry and advanced electronic packaging industry.

Finally, we would like to express our sincere gratitude to our colleagues who have worked hard in the preparation of this series. We also express our heartfelt thanks to those who participated in organizing the publication of this series!

C.P. Wong
IEEE Fellow
Member of Academy of Engineering of the USA
Member of Chinese Academy of Engineering
Former Bell Labs Fellow
Dean of Engineering, The Chinese University of Hong Kong
Regents' Professor, Georgia Institute of Technology, Atlanta, GA
30332, USA

Sheng Liu, Ph.D.
IEEE Fellow, ASME Fellow
Chang Jiang Scholar Professor
Dean, School of Power and Mechanical Engineering
Founding Executive Director, Institute of Technological Sciences
Associate Dean of School of Microelectronics, Wuhan University
Professor of School of Mechanical Science and Engineering
Huazhong University of Science and Technology
Wuhan, Hubei, China

Wenhui Zhu, Ph.D.
National Invited Professor
College of Mechanical and Electrical Engineering
Central South University
Changsha, Hunan, China

Contents

Preface by Yufeng Jin	ix
Preface by Shenglin Ma	xi
Acknowledgments	xv
About the authors	xvii

1 Introduction to HR–Si interposer technology — 1
  1.1 Background — 1
  1.2 3D RF heterogeneous integration scheme — 2
  1.3 HR–Si interposer technology — 7
  1.4 TGV interposer technology — 16
  1.5 Summary — 23
  1.6 Main work of this book — 24
  References — 25

2 Design, process, and electrical verification of HR–Si interposer for 3D heterogeneous RF integration — 27
  2.1 Introduction — 27
  2.2 Design and fabrication process of HR–Si TSV interposer — 31
  2.3 Design and analysis of RF transmission structure built on HR–Si TSV interposer — 38
  2.4 Research on HR–Si TSV interposer fabrication process — 43
    2.4.1 Double-sided deep reactive ion etching (DRIE) to open HR–Si TSV — 43
    2.4.2 Thermal oxidation to form firm insulation layer — 44
    2.4.3 Patterned Cu electroplating to achieve metallization and establish RDL layer — 45
    2.4.4 Electroless nickel electroless palladium immersion gold (ENEPIG) — 54
    2.4.5 Surface passivation — 54
  2.5 Electrical characteristics analysis of transmission structure on HR–Si TSV interposer — 55
  2.6 Conclusion — 61
  References — 63

3 Design, verification, and optimization of novel 3D RF TSV based on HR–Si interposer — 65
3.1 Introduction — 65
3.2 HR–Si TSV-based coaxial-like transmission structure — 69
3.3 Redundant RF TSV transmission structure — 70
3.4 Sample processing and test result analysis — 72
3.5 Optimization of HR–Si TSV interposer — 83
3.6 Conclusion — 90
References — 93

4 HR–Si TSV integrated inductor — 95
4.1 Introduction — 95
4.2 HR–Si TSV interposer integrated planar inductor — 96
4.3 Research on 3D inductor based on HR–Si interposer — 113
4.4 Summary — 123
References — 123

5 Verification of 2.5D/3D heterogeneous RF integration of HR–Si interposer — 125
5.1 Introduction — 125
5.2 Four-channel 2.5D heterogeneous integrated L-band receiver — 126
5.3 3D heterogeneous integrated channelized frequency conversion receiver based on HR–Si interposer — 132
    5.3.1 HR–Si interposer integrated microstrip interdigital filter — 134
    5.3.2 Design, fabrication, and test of HR–Si interposer — 142
    5.3.3 3D heterogeneous integrated assembly and test — 145
5.4 Conclusions — 150
References — 151

6 HR–Si interposer embedded microchannel — 153
6.1 Introduction — 153
6.2 Design of a HR–Si interposer embedded microchannel — 158
6.3 Thermal characteristics analysis of a TSV interposer embedded microchannel — 161
    6.3.1 Simplified calculation based on a variable diffusion angle — 162

6.3.2 Direct calculation based on analytical formula	163
6.3.3 A fitting formula based on simulation results	164
6.3.4 Equivalent thermal resistance network based on the high thermal conductivity path	164
6.4 Process development of a TSV interposer embedded microchannel	172
6.5 Characterization of cooling capacity of HR-Si interposer with an embedded microchannel	176
6.6 Evaluation of HR-Si interposer embedded with a cooling microchannel	178
6.7 Application verification of HR-Si interposer embedded with microchannel	188
6.8 Conclusions	191
References	192

7 Patch antenna in stacked HR-Si interposers — 197

- 7.1 Introduction — 197
- 7.2 Theoretical basis of patch antenna — 200
- 7.3 Design of a patch antenna in stacked HR-Si interposers — 200
- 7.4 Processing of a patch antenna in stacked HR-Si interposers — 213
- 7.5 Test and analysis of patch antenna in stacked HR-Si TSV interposer — 213
- 7.6 Summary — 222
- References — 222

8 Through glass via technology — 225

- 8.1 Introduction — 225
- 8.2 TGV fabrication — 225
- 8.3 Metallization of TGV — 228
- 8.4 Passive devices based on TGV technology — 230
  - 8.4.1 Technology description — 230
  - 8.4.2 MIM capacitor — 230
  - 8.4.3 TGV-based bandpass filter — 231
- 8.5 Embedded glass fan-out wafer-level package technology — 235
  - 8.5.1 Technology description — 235
  - 8.5.2 AIP enabled by eGFO package technology — 236
  - 8.5.3 3D RF integration enabled by eGFO package technology — 242

    8.6 2.5D heterogeneous integrated L–band receiver based on
        TGV interposer   242
    8.7 Conclusions   249
    References   250

9   Conclusion and outlook   **251**

Appendix 1 Abbreviations   255
Appendix 2 Nomenclature   259
Appendix 3 Conversion factors   267
Index   269

Preface by Yufeng Jin

With the rapid development of technologies such as big data, the Internet of Things, 5G communications, smart equipment, and mobile terminals, the development of advanced integrated packaging technologies with higher speed, smaller size, more functions, and lower costs has become a major technical challenge. CMOS SOC technology has entered the 3-nm node, and IBM announced the first 2-nm manufacturing technology on May 6, 2021. However, as the process enters the nanometer or atomic-level era, integration becomes more and more difficult, and the investment required is beyond imagination. New materials, new processes, and new integration architectures are urgently needed to solve the problem of continuous improvement in integration.

In recent years, three-dimensional integration technology has developed rapidly. The three-dimensional integration of semiconductor chips, modules, or systems can greatly increase the integration density and become an effective expansion or extension of Moore's law. Among them, TSV three-dimensional integration technology has outstanding performance. Due to its high integration density and wide application range, it can achieve large-scale production at multiple levels such as chips and modules, and it has become a research frontier in industry and academia. This technology has been widely used in the fields of CMOS image sensors, MEMS devices, RF devices and modules, LED modules, memory stacked modules, and even CMOS SOC.

Worldwide, Georgia Institute of Technology in the United States, IMEC in Belgium, Fraunhofer in Germany, IME in Singapore, Samsung in South Korea, TSMC, ST Microelectronics, and others attach great importance to, and vigorously promote, TSV integration capabilities, and new technologies and products continue to emerge. Similar to the development history of many new technologies, China has also made arrangements in this technical field. The 02 National Major Science and Technology Project, 973, and other programs have successively funded TSV-related technical research. China Aerospace Science and Technology Corporation and Huawei are also paying more and more attention and continuing to invest. However, in general, the foundations in this field are relatively weak, and the cultivating of relevant talents and core technologies is slow to progress.

As the earliest explorer of TSV basic technology, Dr. Shenglin Ma has accumulated rich experience after more than 10 years of painstaking research. He independently completed the development of the first four-layer chip stacked by the TSV process in China during his doctoral dissertation stage. After working at Xiamen University, he conducted much with other superior teams focused on 3D RF integration based on TSV interposer technology, including low-loss RF TSV interposer and the 5–10 GHz 3D integrated channelized frequency converter receiver. Some of the results have been applied in engineering fields such as IPD, TSV three-dimensional radio frequency integration and microchannel heat dissipation, and 38-GHz TSV interposer based antennas. On this basis, the author combined the current developments in TSV RF integration technology and wrote this book: "TSV 3D RF Integration: HR - Si Interposer Technology".

As the author's doctoral tutor, I am very happy to see the publication and write the preface of this book and hope that its publication will play a positive role in promoting the development of TSV technology. As a partner of the author for many years of collaborative research, I was fortunate to read and review all the manuscripts in advance. This book reviews the cutting-edge development of HR-Si interposer technology and introduces details of design and process development of HR-Si RF TSV, the design and process of a new 3D interconnection structure based on HR-Si TSV, the verification of 2.5D/3D RF heterogeneous integration applications, and HR-Si interposer integrated inductor and patch antenna. Furthermore, it also introduces the HR-Si interposer with embedded microchannels. Most of the content of this book is the result of author's accumulated experience and results of research and development in recent years. The experimental and theoretical analysis are closely integrated, and the content is detailed and clear.

I believe that this volume will be of great reference value for scientific researchers, engineering technicians, and graduate students in related fields.

Yufeng Jin
Peking University
Beijing, China

Preface by Shenglin Ma

Since the invention of the transistor, the integrated circuit (IC) industry has continuously developed in line with Moore's law, by which the transistor's feature size has halved every 18 months. The size reduction also results in a cost reduction per transistor, realizing a higher level of integration density and higher energy efficiency. However, as Moore's law is approaching the physical limit, the sharp increase in cost has seriously affected the marginal revenue increase brought by scaling down the transistor's feature size. To maintain or extend Moore's law, new types of devices and materials, including gate-all around (GAA) transistors, carbon nanotube technology, Ge channels, and III–V compound semiconductors, have been proposed and investigated. Although these devices and materials have led to an improvement in integration level and functional complexity, challenges still lie ahead. At the same time, compared to the transistor in ICs whose feature size is continuously being scaled down, advancements in the size reduction and integration density of microelectronics devices like analog and radio frequency (RF) integrated circuits and MEMS/NEMS devices have been lagging behind, as have the line size and precision of the package substrate and assembly substrate, which have become the main restriction factors for system overall performance.

With this as background, 3D heterogeneous integration of CMOS with analog/RF devices and MEMS/NEMS devices, which can improve the comprehensive system performance while elevating the integration level and complexity of functions, has become a key direction of concern within the industry. TSV 3D integration technology can provide a three-dimensional electrical interconnect running through the chip substrate and realize high-density 3D integration through multilayer chip stacking. It is considered an important enabling technology for transcending Moore's law and is attracting extensive attention. At present, TSV 3D integration technology has been commercialized in various fields, including MEMS, CMOS imaging sensors, IR FPA, FPGA, memory ICs, and SOCs, and an accelerating permeation and expansion trend can be seen. An important application direction of this technology is the RF field, in the development of next-generation mobile communication networks and advanced electronic information systems.

TSV interposer technology was first commercialized in 2011 by the US company Xilinx in FPGA devices, and its technical and commercial value are now being widely researched and recognized. Its application in the RF field originated in 2008, when the Institute of Microelectronics in Singapore proposed the concept of TSV interposer-based 3D RF integration technology. Currently, whether in the advanced electronics information system field or the mobile communications field, system-level integrated packaging technology is the main integrated packaging method used to implement RF systems. High-performance package substrate serves as the common interconnected substrate; radio-frequency microelectronics chips/components are mounted on the substrate surface in the form of multichip modules (MCMs), and the electromagnetic shielding structure and plastic package are designed according to the application scenarios. As RF systems develop toward higher frequency, more frequency bands, higher integration, multifunctionalization, microsystems, and lower cost, the system-level package, whether based on an organic package substrate or high-performance ceramic interconnect substrate, is confronted with greater challenges, which on the one hand derive from high-frequency loss, width, line spacing, and precision of interconnecting lines, and on the other hand from high-quality passive elements. The traditional TSV interposer applied to digital ICs is mostly fabricated on a low-resistance Si substrate, and when it is directly applied to high-frequency conditions, high-frequency loss must be higher. Meanwhile, chips in the RF field are of diversified textures. With this background, the HR-Si interposer and glass TGV interposer technologies are promising candidates for integrated package substrates in RF systems and have become research hotspots in this field, largely because they can provide high-precision wiring and low mismatching with integrated microelectronics chips in terms of thermal expansion coefficients, and can provide high-quality passive devices enabled by the use of the MEMS process.

This book presents the preliminary research accomplishments achieved by our research group over the past 10 years with respect to the high-resistivity (HR) Si interposer applied to 3D heterogeneous RF integration. This includes fundamental research on the HR-Si interposer design and process validation, which aims at addressing the core problem of high-frequency loss, new 3D RF interconnection structure design and process validation based on the HR-Si interposer, and application and verification with 3D RF heterogeneous integration to demonstrate its feasibility and technical advantages. In order to elevate the integration level, an HR-Si interposer integrated inductor and integrated patch antenna on stacked HR-Si interposer were studied. Given the heat dissipation problem of high-performance TR integration, an HR-Si interposer with an embedded microchannel was investigated and validated with a 2–6 GHz GaN HEMT-based power amplifier. TGV interposer technology is also reviewed along with research and development advancements in TGV formation, metallization, TGV-enabled IPDs, and glass-based embedded wafer level packages. TGV interposer-based 2.5D/3D RF integration by Professor Daquan Yu and my team is presented, to provide a full view of interposer technology for 3D RF integration.

Several books in English are available to readers on the "more than Moore" (MtM) electronics packaging. However, most of them are limited to the conceptualization of integration or package architecture, electrical, and process evaluation. This book focuses on a specific topic: a 3D heterogeneous RF integration oriented HR-Si and TGV interposer from our research, with the intention of reviewing the progress in this field to communicate with researchers who have interests or will be engaged in the related technology areas. Constructive suggestions or criticisms are appreciated, in hopes of putting forward the advances in this field.

Since it is co-published with Elsevier Inc., the present edition follows the typesetting of Elsevier's edition, including, but not limited to, the fonts, sizes, subscripts, superscripts, normal or italic letters, reference, as a courtesy. Metric units have been used throughout this book, though a few common imperial units appear only on very limited occasions. The readers can refer to Appendix 3 for the conversion factors between imperial units and metric units.

This book is intended for design engineers, processing engineers, and application engineers as well as for postgraduate students majoring in advanced electronics packaging.

Shenglin Ma
Xiamen University
National Key Laboratory of Science and Technology
on Micro/Nano Fabrication
Peking University

Acknowledgments

Thanks to my PhD supervisor, Professor Yufeng Jin from Peking University; coauthor of this book: Professors Wei Wang, Min Miao, and Jin Chen; my partners Liulin Hu and Shuwei He from Chengdu Ganide Technology Corporation Ltd.; my graduate students Jun Yan, Han Cai, Yunheng Sun, Mengcheng Wang, Tingting Lian, Yang Yang, Yanming Xia, Kuili Ren, Luming Chen, Juan Yu, and Nanxin Wang; research assistants Honglin Gong, Zhizhen Wang, and Lili Wei; and Editor Gang Wu from Chemistry Industry Press Co., Ltd., China. Without them, it's unimaginable that this book could have been completed. I am grateful to them all, and I wish them a bright future.

A special thanks to my friend Professor Daquan Yu, an expert in both industry and academics, who now is engaged in the commercialization of TGV technology. His excellent work in this field is presented in Chapter 8. Also, thanks go to his team members Tian Yu and Li Chen, who took part in the drafting of the chapter.

I am very grateful to my partners, my family, and my friends for their unfailing encouragement and support.

Shenglin Ma, Ph.D., Associate Professor
Xiamen University
National Key Laboratory of Science and Technology on Micro / Nano Fabrication,
Peking University
Beijing, China

About the authors

Shenglin Ma received MS and PhD degrees in microelectronics engineering from Peking University, Beijing, China, in 2008 and 2012, respectively. Since 2012, he has been working for the Department of Mechanical and Electrical Engineering, Xiamen University, Xiamen, Fujian, 361005, China, and for the National Key Laboratory of Science and Technology on Micro/Nano Fabrication, Peking University, as guest researcher. He has taken part in 10 projects as a key researcher, has had more than 50 publications in international journals and conferences, more than 20 patents granted in China, and 3 patents granted in the United States. His research interests include TSV-based 3D integration technology, MEMS packages, and reliability.

Yufeng Jin received his PhD degree in physical and optical-electronics engineering from Southeast University, China, in March 1999. Since then, he has worked as a postdoctoral fellow, associate professor, and professor at Peking University. He has directed the National Key Laboratory of Science and Technology on Micro/Nano Fabrication since 2005. His research interests focus on MEMS sensors, TSV-related 3D integration of microsystems, and their applications.

CHAPTER 1

Introduction to HR-Si interposer technology

1.1 Background

Since the invention of the transistor, the integrated circuit (IC) industry has continuously developed by following Moore's law, by which the transistor's feature size has been halved every 18 months. The size reduction also indicates a cost reduction per transistor, realizing a higher level of integration density and higher energy efficiency. However, as Moore's law is approaching the physical limit, the sharp increase in cost has seriously affected the marginal revenue increase brought by scaling down the transistor's feature size. To keep Moore's law valid, new types of devices and materials, including gate-all around (GAA) transistors, carbon nanotube technology, Ge channels, and III-V compound semiconductors, have been proposed and investigated. Although these devices and materials have led to an improvement in integration level and functional complexity, challenges still lie ahead [1–7]. Relative to the transistor in ICs, advancements in the size reduction and the integration density of other microelectronic devices, such as analog and radio frequency (RF) components and microelectromechanical/nanoelectromechanical system (MEMS/NEMS) devices, lag behind. The size and the relative precision of the redistribution layer of package substrate and assembly substrate also lag behind. So, the system overall performance is restricted.

With this background, the CMOS IC has been integrated with analog/RF devices, MEMS/NEMS devices, and other chip-level 3D devices. This can improve the comprehensive system performance while elevating the integration level and complexity of functions [8,9], thereby becoming a key direction within the industry, as is acknowledged increasingly. Through silicon via (TSV) 3D integration technology can provide the 3D electrical interconnect running through the chip substrate and realize the high-density 3D interconnects through the multilayer chip stacking. This technology has been considered an important enabling technology for continuing or transcending Moore's law, and has attracted extensive attention. At present, the TSV 3D integration technology has been commercialized in various fields, such as MEMS, CMOS imaging sensors, infrared focal plane arrays, field-programmable grid arrays, memory ICs, and SOCs, and is demonstrating an accelerated permeation and expansion trend. An important application direction of this technology is the RF field under the development of next-generation mobile communication networks and advanced electronic information systems.

Whether in advanced electronic information systems or mobile communications, the system-level packaging technology is mainly used to implement the RF system. The high-performance package substrate, including microwave printed board, low temperature co-fired ceramic (LTCC), and high temperature co-fired ceramic (HTCC), serves as the common interconnection substrate. RF/microwave microelectronic chip/components are mounted on the substrate surface in the form of multichip modules. The electromagnetic shielding structure and plastic package are designed in accordance with the specified application scenarios. As the RF front-end system is developing toward higher frequency, more frequency bands, high integration, multifunctionalization, microsystems, and low cost, the system-level package is faced with increasing challenges. On the one hand, these challenges are derived from the high-frequency loss, width, line spacing, and precision of interconnecting lines. On the other hand, they are derived from high-quality passive elements, and the two jointly decide the system volume and size, integration level, and performance. The traditional TSV interposer applied to digital ICs is mostly fabricated on a low-resistivity Si substrate. When it is directly applied to the high-frequency field, high-frequency loss is confronted. In addition, the chips are made of diversified textures in the RF field.

With this background, the high-resistivity Si (HR-Si) interposer and through glass via (TGV) interposer technologies have been proposed to be advantageous competitors for package substrates in RF systems, because they can address the high-frequency loss issue, provide high-precision wiring, a low mismatch with integrated microelectronic chips in terms of thermal expansion coefficient, and high-quality passive elements enabled by the MEMS process.

1.2 3D RF heterogeneous integration scheme

To miniaturize the RF module, and improve the performance and integration level as well, research efforts have been devoted to the transistor level, the chip level, and the module assembly level. In the transistor level, the monolithic microwave integrated circuit (MMIC) has advanced in recent years, especially in GaN technology. To conquer the limit available of transistors in a specified substrate, heterogeneous integration of transistors was introduced, such as lattice engineering, which can realize CMOS devices and III–V devices, such as HBT and HEMT, fabricated side by side on the same substrate, and peeled transistors such as GaN HEMT from various substrates to be transferred to a common substrate by means of chip-wafer bonding. Based on lattice engineering, SOITEC has claimed that their customized SOI wafers (4, 6, 8 in.) are capable of supporting selective growth of monocrystal thin film of GaN, GaAS, and InP. Raytheon has demonstrated the heterogeneous integration of a 1-μm InP HBT device and 180-nm process node CMOS device on a customized 8-in. SOI substrate. However, the quality of III–V monocrystalline thin film remains to be further improved as it is being selectively grown in the presence of NMOS and PMOS devices, and the III–V monocrystalline film and the relative devices need a collaborative design and process development for CMOS devices. While for the transistor transfer method, the performance of the thin transistor layer can be easily affected or degraded, it is still being improved in the comprehensive performance and diversity of integrated devices.

Besides integrating more kinds of transistors on a common substrate, vertical integration of various high-performance transistors atop an active chip or wafers by stacking layers of transistors is another popular method. As shown in Fig. 1.1 [10], the III–V substrate wafer, where the fabrication of transistors is finished while the fabrication of the interconnect of transistors is not completed, is flipped and bonded to the active side of a CMOS wafer, thinned to expose the buried transistors in the III–V substrate, and then the fabrication of interconnection between CMOS top of the III–V based transistors is continued. Fig. 1.1B shows a 250-nm InP heterojunction bipolar transistor (HBT) stacked on a 130-nm RF CMOS chip developed by HRL Laboratories. In this scheme, the transistor level alignment is implemented between stacking transistor layers, which means a high accuracy requirement, and the InP device performance is faced with the risk of compromise due to integration. In a similar manner, more layers of transistors can be stacked. One point to be noted is that the stacked layer of transistors may have finished transistor interconnection; if so, it means that it has changed to chip level integration and the alignment requirement can be reduced to that between the stacked chips. In 2019, researchers from Stanford University in the United States demonstrated a sample of four stacked layers of transistors, including a 1-μm CMOS chip at base, a layer of carbon nanotube field-effect transistors (CNFET), RRAM, and a CNFET sensing device layer.

In recent years, advanced packaging technology enabled by TSV, microbump, and chip or wafer level bonding technology, has inspired new ways to improve the 3D RF integration level in future, and has offered a chance for a performance breakthrough. For example, with the microbump bonding process, the active surfaces of GaN devices, InP-based (HBT), and GaAs devices can be directly bonded on a CMOS chip in the form of chip-to-wafer or wafer-to-wafer. Based on this technology, BAE and Teledyne Scientific Company have demonstrated an ultrawide band (UWB) ADC sample, as shown in Fig. 1.2 [11]. The frequency range of UWB is 2.75–8.75 GHz and 14.25–20.25 GHz, with the most advanced signal-to-noise ratio (over 30 dB) and distortion ratio, demonstrating a revolutionary breakthrough of performance improvement. By introducing TSV to the RF microelecronic chip, 3D RF integration can be easily realized by chip stacking with microbump. Researchers from Teledyne Scientific Company in the United States have demonstrated a stacked chip module with a 250-nm InP HBT standing upon a 150-nm GaN HEMT chip. Fig. 1.3 shows a 3D heterogeneous integrated phased assay module financed by DARPA [12] in the United States, where a glass interposer containing the patch antenna array is stacked on an SiGe beamformer with TSV

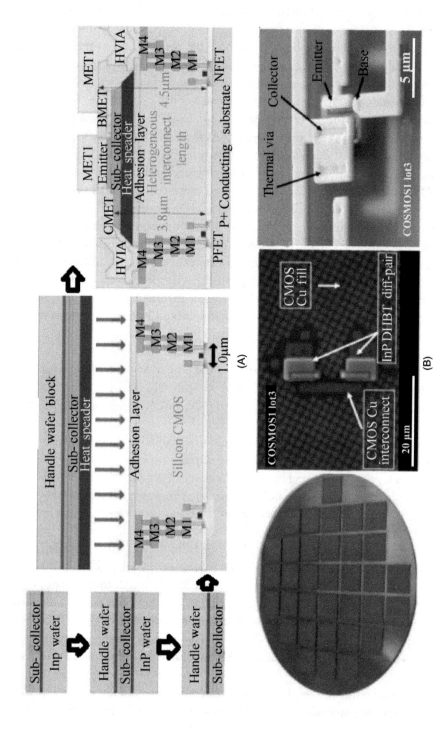

Fig. 1.1 3D heterogeneous integration by stacking transistor layers developed by US HRL Laboratories: (A) process flow and (B) sample picture.

Fig. 1.2 Microphotograph of COSMOS ADC consisting of InP chip and Si CMOS substrate chip.

interconnect (Fig. 1.3A); the cross-section of 3D integrated sample is shown in Fig. 1.3B, and the function test results of amplitude control and phase shift are shown in Fig. 1.3C and D, respectively.

At the assembly level, a 3D multichip module based on advanced ceramic substrate has been developed, in which a double-sided chip assembly on substrate and substrate stacking with soldering are used. To further miniaturize the module, an HR-Si substrate based on a typical MEMS process is proposed to replace the advanced ceramic substrate or even the relative metal housing, to achieve a full Si packaged RF module with small form factor. To realize the electrical connection from inside the Si package to the outside, various methods have been studied, including formation of big through Si holes with a KOH or TMAH etching process to expose the buried input/output electrodes, electromagnetic coupling structure, metallized through-hole by a conformal deposited metal layer on the sidewall of holes, and the introduction of TSV interconnect in recent years. Fig. 1.4 shows a diagram of the HR-Si package-based 3D RF integration module. Some research institutes claim that an on-shelf product based on this technology is available.

Whether transistor level or chip level, collaborative design and process development is needed between the laminated chips in any 3D heterogeneous integration scheme as stated previously, indicating great constraints must be confronted. The transistor level is of a high integration level, but its current maturity level is low due to the process of the ultrathin device layer. Considering the alignment precision requirements and known good die yield loss, this level presents great technical difficulties and high risk. The chip level is preferred due to the moderate process requirement and technical advantages in terms of form factor, integration level, and performance. Nowadays, the HR-Si based multichip module is more than a chip-scale package solution for RF microelectronic chips; the technical advances or advantages of HR-Si substrate, especially enabled by TSV technology, are worthy of major expectations, and it should go beyond as a substrate with horizontal RF transmission, or vertical power or digital signal transmission, and develop into an independent functional unit in the RF module.

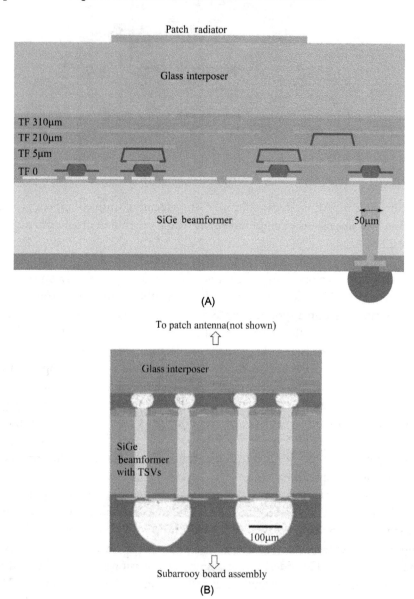

Fig. 1.3 Concept map of (A) 3D integrated phased array cross section, (B) subarray component developed by DARPA

(Continued)

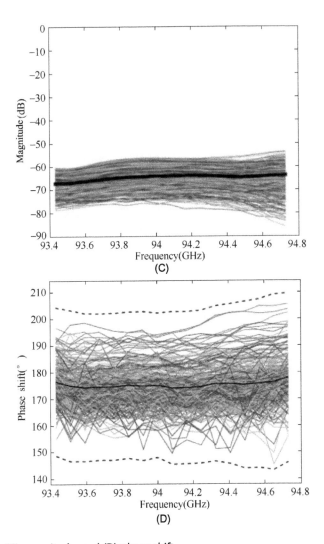

Fig. 1.3, cont'd (C) magnitude, and (D) phase shift.

1.3 HR-Si interposer technology

The HR-Si interposer is basically composed of TSVs and rewiring lines, working as a common interconnect substrate to form an open 3D integration structure. From the view of fabrication, the HR-Si interposer can be implemented with 3D RF integration indivisibly or independently. Fig. 1.5 shows an HR-Si interposer based 3D integration scheme. At first, an HR-Si substrate with blind TSV and cavity is formed, and then RF microelectronic chips, such as GaN HEMT, are assembled inside the cavities and the rewiring layer is fabricated upon them to interconnect the chips and provide passive

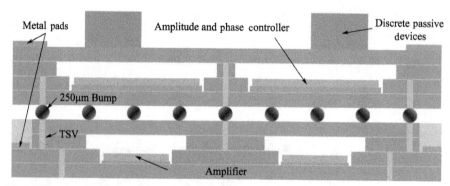

Fig. 1.4 A diagram of HR-Si package–based 3D RF integration module.

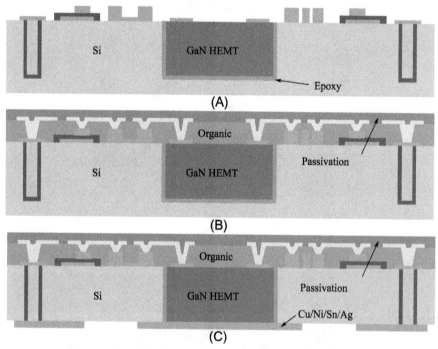

Fig. 1.5 Embedded HR-Si wafer-level heterogeneous integrated packaging technology.

devices at the same time. The 3D vertical integration can be further implemented by TSV and microbump [13].

In addition, more studies have been done with the HR-Si interposer being taken as an independent part in 2.5/3D RF integration in recent years, such as in research from French CEA-Leti, Silex, IMEC, Peking University, Xiamen University, and the 55th, 13th, and 29th Research Institutes of China Electronics Technology Group

Corporation [12–18]. This is more likely preferred by Outsourced Semiconductor Assembly and Testing (OSAT) or pure MEMS foundries. The representative research results include the HR-Si interposer-based 3D integrated T/R module reported by French CEA-Leti at the Electronic Components and Technology Conference (ECTC) in 2015 and 2019 [16], and the HR-Si interposer-based 2.5D heterogeneous integrated L band receiver disclosed by Xiamen University at the IEEE International Conference on Integrated Circuits, Technology and Applications in 2018 [17].

In 2015, researchers from CEA-Leti reported the TSV-last process-based HR-Si interposer technology [16]. The process is as follows. First, two copper wiring layers are fabricated using an IC back-end metallization process on the HR-Si substrate. Second, blind holes are etched on the back face of the HR-Si substrate until the buried first metal wiring layer is exposed. An oxide layer is formed on the side wall of the TSV blind hole, and a window is opened at the hole bottom. Finally, a continuing copper layer is formed inside the TSV blind hole via the copper electroplating process. The diameter and depth of the TSV hole are approximately 60 and 110 μm, respectively. The thickness of the continuous copper layer in the TSV hole is $\leqslant 10$ μm. The HR-Si interposer-based 2.5D integrated TR module proposed by CEA-Leti is shown in Fig. 1.6. According to the test results of the CPW line linked TSV sample, the insertion loss of TSV interconnect was smaller than 0.45 dB within 0–60 GHz. Fig. 1.7 shows the optical microphotographs of the HR-Si interposer-based 2.5D integrated TR module. Fig. 1.7A is the photo of the HR-Si interposer, and Fig. 1.7B and C show the photos of a 65-nm CMOS RF IC transceiver chip being flipped chip bonded on the HR-Si interposer with molding and soldering, viewed from bottom side and topside, respectively. The module is assembled on the printed circuit board (PCB) through solder balls, thereby forming a millimeter wave TR module with dimensions of $6.5 \times 6.5 \times 0.6$ mm^3. The measured reflection coefficients of two different antenna samples, E3 and E4 in Fig. 1.8A, are kept below −10 dB within the frequency band of 57–67 GHz (VSWR < 2), corresponding to the

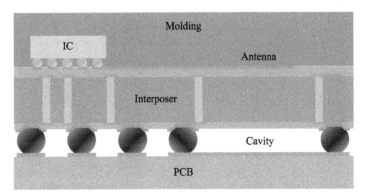

Fig. 1.6 The HR-Si interposer-based 2.5D integration concept proposed by CEA-Leti.

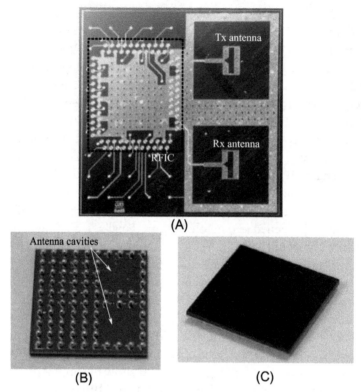

Fig. 1.7 Optical microphotographs during the module assembly process of HR-TSV interposer.

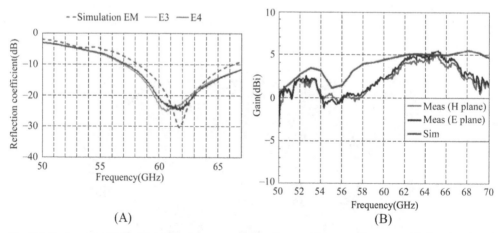

Fig. 1.8 Test results (A) Simulated and measured reflection coefficients of antenna. (B) Simulated and measured gains (dBi) of antenna (H and E planes).

bandwidth of 16%. The gains measured on the H and E planes of the antenna are shown in Fig. 1.8B [18], where the gain at 60 GHz is 2.3 dBi, and the maximum gain (5.55) is acquired at 65.5 GHz. This module can realize the 3.5-Gbps communication function within the range of 1–3 m. The basic functions of this millimeter-wave TR module have been experimentally verified.

In 2019, researchers from French CEA-Leti updated the progress on HR-Si interposer-based 2.5/3D integration technology. Integration of an inductor on the HR-Si interposer was fabricated and tested. A dual band TR module was designed, and the overall area was reduced by approximately 60% in comparison with the counterpart implemented with PCB. This study reported measurement results of multifrequency multimode power amplifiers with operating frequency of 400 and 900 MHz. In the low band, the test shows output power performances of about 28 and 29 dBm working in 3GP mode and PMR mode, respectively. While in the high band, the test shows output power performances of about 28.5 and 30 dBm working in 3GP mode and PMR mode, respectively. For the interference of the long-term evolution (LTE) system with the other LTE systems, the output power in the adjacent channel leakage power ratio test was −37 dBc, but the overall functional test of the TR module was not completed by the time of paper publication. The HR-Si interposer for the 3D integrated TR module is shown in Fig. 1.9. The representative study results are shown in Table 1.1.

In recent years, studies of the HR-Si TSV-based integrated passive devices, including capacitor, inductor, and filter, have achieved rapid progress. The representative study will be elaborated as follows.

In 2013, Swiss Selix displayed the application of RF TSV-based 3D IPDs [21] by the design and manufacture of a group of inductors with HR-Si interposer. The inductance value ranged from 1.38 to 11.91 nH, while the peak Q value ranged from 134 to 14, with self-resonance frequency of >10 GHz. Efforts have been exerted to repress the substrate coupling effect by using low-loss tangent materials and low dielectric constant materials for improving the inductor quality factor. In the literature [19], an air bridge was used

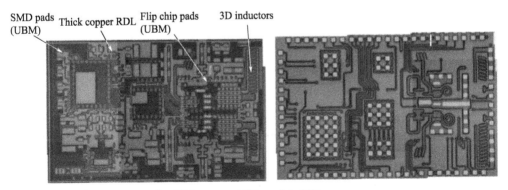

Fig. 1.9 The HR-Si interposer for 3D integrated TR module [19].

Table 1.1 Representative study of HR-Si interposer technology-based 3D RF integration.

References	Material/ structure/ feature size	Demonstrator	Tested technical parameters	Target application
SELIX [15] (2013)	Diameter: 200 mm Thickness of Si substrate: 300 μm TSV diameter: 90 and 50 μm	N/A	TSV: S21 < 0.04 dB at 5 GHz DC resistance: 20 mΩ Integrated inductor: a Max. inductance value of 5 nH, a peak Q value of 30, resonant frequency no less than 10 GHz	RF 3D integration, the inductor is applicable to RF and power SOC
CEA-LETI [12] (2013)	Thickness of Si substrate: 120 μm TSV diameter: 60 μm, depth-to-width ratio: 2:1 UBM solder pad: Ti/Ni/Au, diameter: 290–325 μm Solder ball: 80 μm	A chip scale package TR module	TSV: S21 < 0.45 dB at 60 GHz Antenna bandwidth range of −10 dB is 57–67 GHz. 65.5 GHz gain is 5.55 dBi	Applied to 60-GHz fast data transmission
CEA-LETI [20] (2019)	Thickness of Si substrate: 200 μm RDL thickness: 7 μm	3D integrated TR module	DC resistance of RDL: 65 mΩ/mm DC resistance of Cu TSV: 3–4 mΩ	Multiband, multimode TR module: 400 and 900 MHz

above the benzocyclobutene dielectric medium as the interface layer of the inductor on the lossy Si interposer. The peak Q value, inductance value, and self-resonance frequency are measured as 22 at 24 GHz, 0.83 nH, and 59 GHz, respectively. Fig. 1.10 shows the 2D planner inductor and 3D inductor fabricated in an HR-Si interposer by researchers from French CEA-Leti. The maximum quality factor of the 2D planner spiral inductor is 25 at 2 nH and that of the 3D inductor can reach 35 at 2 nH, as shown in Fig. 1.11. It can be

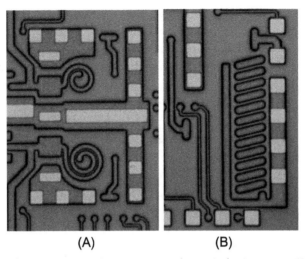

Fig. 1.10 Views of inductors in an HR-Si interposer under optical microscope: (A) 2D planner spiral inductor on back face and (B) 3D inductor on back face.

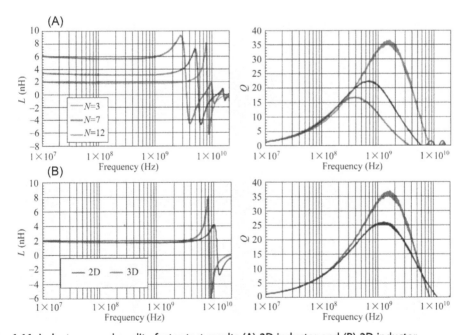

Fig. 1.11 Inductance and quality factor test results (A) 3D inductor and (B) 2D inductor.

found that the TSV-based 3D inductor is helpful to increase the inductance value and quality factor.

In 2012, the Institute of Microelectronics of the Chinese Academy of Sciences developed a TSV array PN junction capacitor [20]; a capacitance density of 2 nF/mm^2 was obtained, showing potential application in a voltage range from 0 to −7 V. In 2014, CEA-Leti presented a new type of TSC capacitor structure [22] (Fig. 1.12), the capacitance density reaching 23 nF/mm^2 and having a low parasitic equivalent series resistance and inductance, and high series resonance frequency (about 10 GHz when the capacitance value is 10 nF).

For the integrated filter on a TSV interposer, Xidian University designed and processed a planar LC filter. The measured cut-off frequency was 10.05 GHz, the insertion loss in the passband was 0.14 dB, the reflection loss exceeded 13 dB, and the inhibition level was higher than 20 dB within the stop band. ON semiconductor developed a 5G n77 bandpass filter (BPF) based on the TSV process [23], as shown in Fig. 1.13.

The antenna size is substantially reduced when the operating frequency enters the millimeter wave band, making the antenna integration possible. The high dielectric constant of Si material goes against the improvement of the antenna emission ratio. However, Si materials perform well in terms of manufacturability and process compatibility. In 2012, CEA-Leti developed a 60-GHz cavity-backed antenna array integrated on the

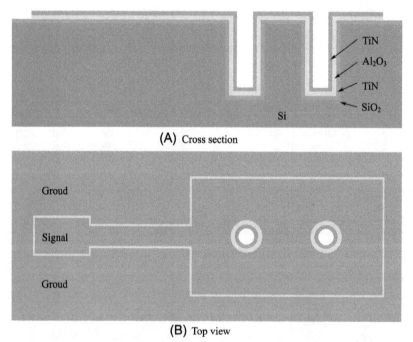

Fig. 1.12 Schematic of two-TSC matrix capacitor: (A) cross-section view and (B) top view.

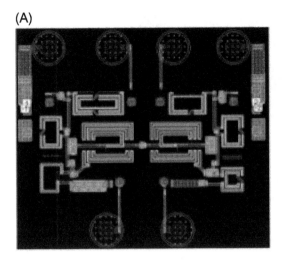

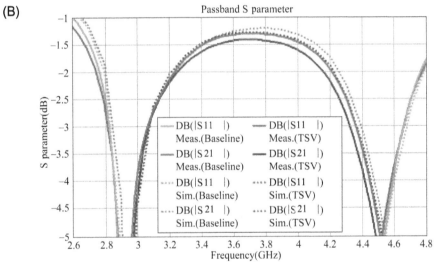

Fig. 1.13 An N77 filter based on TSV technology: (A) photo views from both sides and (B) insertion loss test results.

HR-Si interposer, as shown in Fig. 1.14. The experimental results showed that this antenna worked in the range of 57–66 GHz, with a gain exceeding 5 dBi, as shown in Fig. 1.14 [16]. In 2019, the 55th Research Institute of China Electronics Technology Group Corporation proposed a compact-type planar doublet antenna on a Si substrate. The simulation results showed that the operating frequency range with the return loss smaller than −10 dB was 89.5–101 GHz, and the relative bandwidth was 12.07%. Within the whole operating frequency band, a high antenna gain (4.7–6.03 dBi) could be acquired.

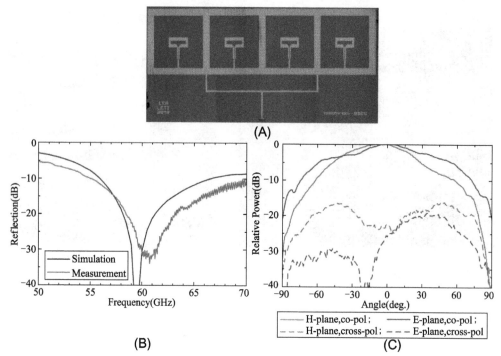

Fig. 1.14 A folded dipoler antenna integrated in HR-Si interposer: (A) sample, (B) reflection, and (C) radiation profile.

1.4 TGV interposer technology

Glass material is an insulator, with a dielectric constant about one-third that of silicon, and the loss factor is two to three orders of magnitude lower than that of silicon. So, the substrate loss and parasitic effects are greatly reduced, and the high-frequency electrical performance is excellent. Large-size and ultrathin glass substrates are available commercially. Glass manufacturers such as Corning, Asahi, and SCHOTT can provide ultralarge (>2 m × 2 m) or ultrathin (<50 μm) panel glass. Therefore glass is proposed to replace lossy Si in a traditional TSV interposer. The TGV interposer does not require an insulating layer, making it a simple process. It can integrate passive devices on itself and can fully inherit and utilize the accumulation of manufacturing technologies such as glass substrates and printed circuit boards to improve production efficiency. Besides application to the high-frequency field, glass is a transparent material, so a TGV interposer can be used in the fields of optoelectronics and MEMS. In recent years, institutions researching TGV have included Corning in the United States, SCHOTT in Europe, PlanOptik AG in Europe, NEC SCHOTT, AGC in Japan, Georgia Institute of Technology in the United States, Fraunhofer IZM in Germany, Singapore Institute of Microelectronics (IME), Korea Industrial Technology Research Institute, Japan Industrial Technology

Research Institute, RTI in the United States, and other international leading glass material suppliers and semiconductor research institutions.

One of the difficulties in the development of a TGV interposer is the TGV formation. Over the years, many research efforts have been devoted to finding low-cost, small-size, fine-pitch, and nondestructive rapid hole-forming technology. The fragility and chemical inertness of glass materials make it more difficult. The main TGV-forming methods include sandblasting, laser ablation, photosensitive glass, focused discharge, and laser-induced etching. Among them, the sandblasting method is a hole-forming technology first applied in the field of MEMS. The diameter of the sand particles used in the process is 20–50 μm, and the via size and spacing are relatively large. The damage to the glass surface and the side wall of the hole introduced during the via formation is the main problem.

Laser ablation mainly uses femtosecond laser, picosecond laser, nanosecond excimer laser, and CO_2 laser. The CO_2 laser drilling process is mainly formed by the partial melting of the glass material by the thermal effect. It belongs to the thermal laser category, which has many sidewall cracks and thermal stress problems. The excimer laser belongs to the "cold laser" category, and the formed hole wall basically has no cracks. The via-forming efficiency is not high. Fig. 1.15A is a group of vias formed by laser cauterization. Photosensitive glass is combined with photolithography to form a hole. The pattern is defined by ultraviolet irradiation, and then through a high-temperature sintering process, the material properties of the area irradiated by ultraviolet light are transformed into ceramic materials. Finally, the ceramic material is removed by HF acid etching. The main technical problem facing the realization of vias having different sizes of patterns such as blind holes or blind grooves is that the corrosion rate is different, which results in a large difference in the accuracy of the pattern definition [24], as shown in Fig. 1.15B. The focused discharge via forming method focuses on high-voltage discharge on the upper and lower sides of the glass, causing the glass to melt and evaporate in a local area to form TGVs. This method can achieve a minimum of 40–60 μm TGV holes on 300 μm and 500 μm thick glass as shown in Fig. 1.15C. The time for a single via is 200–500 μs.

The plasma etching method can form holes in parallel on the wafers, with small sidewall roughness (<150 nm), sidewall erroneous damage, and good reliability guarantee. At present, the main technical difficulty is the complexity of the process. The etching rate is usually less than 1 μm/min.

The laser-induced etching method uses laser pulses to induce the glass to produce denatured areas. The deformed area and the surrounding substrate have a high corrosion selection ratio in diluted HF acid. Based on this principle, a continuous denatured area is formed by laser, and it is transformed by etching. The research results of LPKF show that the laser-induced hole formation method can be applied to common glasses provided by a variety of manufacturers, and can form 10 to 50 μm TGV vias on 50 to 200 μm thick glass. The hole formation rate is fast, the damage is small, and the laser processing hole

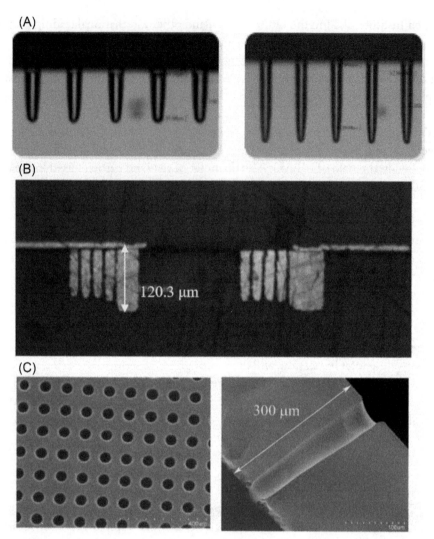

Fig. 1.15 TGVs made by different methods: (A) laser ablation, (B) photosensitive glass, and (C) focused discharge.

frequency can reach 5 kHz, but due to the anisotropy of the etching, the through hole has a certain slope. If it is applied to the production of patterned through holes, the hole-forming speed will decrease. The laser-induced etching method has great application potential.

TGV metallization is another technical difficulty. In the early metallization schemes, tungsten and copper were mainly used. Fig. 1.16 [25] shows the tungsten TGV interconnect glass wafer technology developed by European PlanOptik AG and NEC SCHOTT.

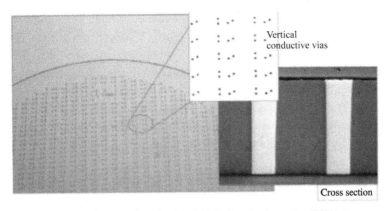

Fig. 1.16 Glass wafer technology with embedded high-density tungsten TGV interconnection.

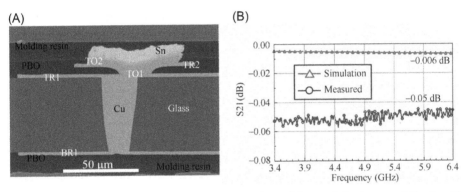

Fig. 1.17 Study results of TGV interposer by Taiwan Industrial Research Institute and Corning Glass: (A) TGV samples and (B) RF transmission performance is arranged with one signal TGV in center circled by six grounding TGVs (1S6G).

Unlike silicon materials, glass has poor adhesion to common metals such as copper, and is prone to delamination, which causes the metal layer to peel off or curl. Before copper electroplating, it is necessary to deposit an adhesion layer such as Ti, Cr, etc., and a seed layer of Cu. An AGC study showed that the adhesion between the Cr layer and ENA 1 glass is the best (347.8 mN), which is greater than the 244.1 mN between the silicon and TiW metal layers [26]. Fig. 1.17A shows the study results of a TGV interposer reported by the Taiwan Industrial Technology Research Institute and Corning Glass. It is similar to the metallization scheme of the traditional TSV interposer, with drill blind vias, PVD seed layer, and bottom-up copper electroplating to achieve seamless filling of TGV. A Cu redistribution layer (RDL) is formed. Then through temporary bonding, back grinding, and CMP, the buried end of the Cu TGVs is continued to be exposed. Finally an RDL is formed in a similar way on the exposed surface and debonded. The solid copper TGV interconnection has a diameter of 100 μm and a height of 300 mm, and two RDL wiring layers on the front and one RDL on the back, which have a line width of 20 μm and a thickness of 4–5 μm. Fig. 1.17B shows the test results of the 1S6G structure. Due to the solid copper filling, the large TGV aperture faces a greater risk of thermal stress reliability. At present, TGV metallization adopts a conformal deposition to realize a hollow Cu or Au TGV interconnection. Due to the skin effect at high frequency, the difference in high-frequency electrical properties between hollow and solid TGV interconnects is small.

Researchers are developing a low-cost filling solution for glass vias. Due to the different thermal expansion coefficients between glass and copper (glass 3 ppm/K, copper 17 ppm/K), there are obvious differences in chemical structure. In addition, the glass has a very smooth surface, which results in poor adhesion between glass and electroless copper. Thus special treatment is necessary to improve the binding force. Atotech proposed that the glass substrate be immersed in a chemical solution to cover the nanothickness metal oxide as an adhesion layer to improve the adhesion of the electroplated copper layer. A picture of the 9-nm thick adhesion layer is shown in Fig. 1.18 [27]. For the 5–20-nm adhesion layer, it can make the peeling strength between copper and glass reach over 6 N/cm. Onitake et al. developed a direct electroless copper plating in combination with a following copper electroplating process with a photoresist pattern mask after UV cleaning based on a 254-nm light wave (KOL1-1200U, Koto Electric Co., Ltd.). Ultraviolet light cleaning technology uses the photosensitive oxidation of organic compounds to remove organic substances adhering to the surface of the material, and the surface of the

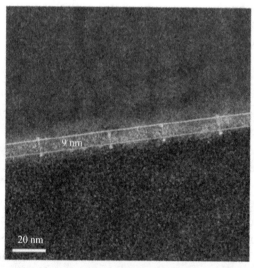

Fig. 1.18 The method of glass substrate immersed in a chemical solution to deposit metal oxide as an adhesion layer proposed by Atotech of the United States.

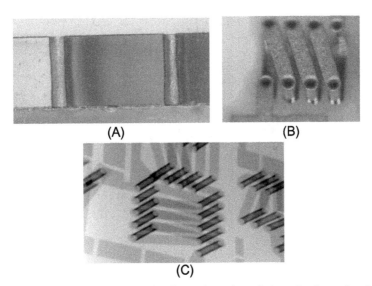

Fig. 1.19 The copper seed layer produced by direct electroless plating after laser cleaning proposed by AGC Company, and the TGV-based inductor sample realized in combination with SAP subsequently: (A) cross-sectional view, (B) top view, (C) X-ray image of view of the sample [29].

material after cleaning can achieve atomic cleanliness [28]. Fig. 1.19A shows the cross-sectional view of TGV finishing electroless copper plating; a continuous Cu layer of 3 μm in thickness is obtained. Fig. 1.19B and C show the fabricated TGV-based inductor. The test results show that the peel strength between copper and glass is 3.5 N/cm.

Georgia Tech researchers used a glass substrate with TGVs as the core layer of a PCB board, laminated epoxy polymer film on the thin glass substrate both sides, and made through holes on the filling medium in the TGV by CO_2 laser. Then, they used SAP to make the rewiring layer, as shown in Fig. 1.20. Table 1.2 shows the specifications of the TGV interposer [29]. The laminated dielectric layer can enhance the mechanical strength of the glass substrate, and, meanwhile, work as the adhesion layer between the metallization layer and the surrounding substrate to enhance the peel strength and reduce the difficulty and cost of TGV metallization. Copper conductive paste or other metal conductive paste is proposed to make a low-cost metallization. However, the electrical performance of the conductive adhesive is poor, which limits its application at high frequency.

Regarding 3DRF integration, there are two main application scenarios for the TGV interposer. One is used as a common substrate for mounting chips made of different materials, and the other is to realize passive devices such as inductors, filters, and antennas [30]. Fig. 1.21 is a 3D heterogeneous integrated RF front-end module demonstrated by Georgia Institute of Technology in the United States, in which a TGV-based filter is implemented; no function test results were disclosed at the time of publication. A joint research team from University of Florida, Corning, and DuPont, etc., reported

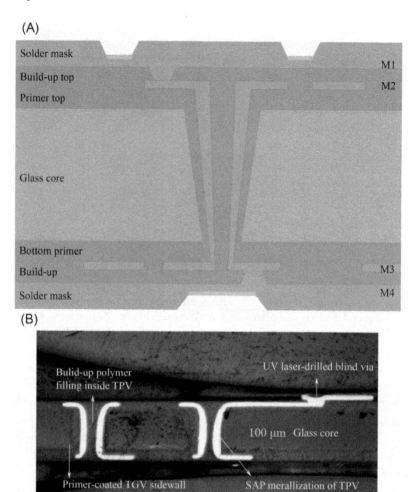

Fig. 1.20 TGV samples made by Georgia Tech through a PCB-like process: (A) schematic view of TGV samples, (B) cross-sectional view of TGV samples.

Table 1.2 Specifications of TGV interposer by Georgia Tech.

Layer	Thickness (μm)	Material
M1	10	Cu
Build up top	15	GX-92
M2	8	Cu
Primer top	15	GX-92
Glass core	100	Glass
Primer bottom	15	GX-92
M3	8	Cu
Build up	15	GX-92
M4	10	Cu

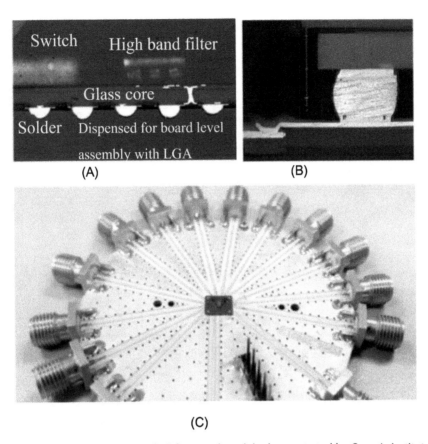

Fig. 1.21 A 3D heterogeneous integrated RF front-end module demonstrated by Georgia Institute of Technology in the United States: (A) cross-sectional view of full-chain module, (B) high-band SAW filter assembly on glass, and (C) full-chain module mounted on evaluation board [30].

a TGV interposer integrated W-band (75–110 GHz) antenna, where TGV is metallized by double-sided TiCu sputtering of 30 nm/2 μm, the return loss at 78 GHz is 21.6 dB, and the 10 dB bandwidth is 70.75–83.67 GHz, which is about 16.6%.

1.5 Summary

In summary, the applications of an HR-Si/TGV interposer to 3D heterogeneous RF integration have been demonstrated in recent years, and its role has been clarified more clearly. First, it provides a high-frequency, low-loss, and high-density interconnected common substrate. Second, it provides high-quality passive devices. The high-frequency loss is still a key issue needing to be addressed carefully. One point that needs to be made is that, in comparison with the traditional passive devices, the IPD element of the TSV/

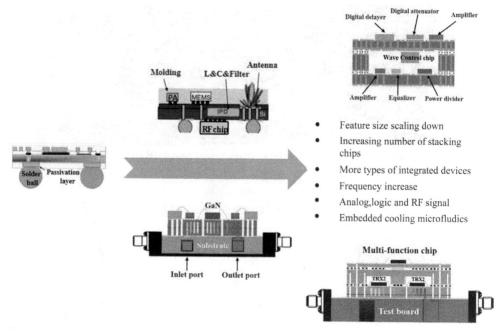

Fig. 1.22 Developmental roadmap of HR-Si/TGV interposer technology.

TGV interposer is developing into a unique, independent, passive device gradually by the virtue of its advantages in size, line width precision, performance and quality, and compatibility with integration. It will play an increasing role in the RF integration system in the future, with good commercialization prospects.

In the future, the HR-Si/TGV interposer will advance in the following aspects: higher frequency applications, more stacking layers, more kinds of integrated chips, integration of digital signals, analog signals, optical signals, and cooling microfluidics, as shown in Fig. 1.22. We hope more peers will contribute their efforts to Chinese chips by further developing applications of HR-Si interposer technology for 3D heterogeneous RF integration.

1.6 Main work of this book

In this book, the preliminary research accomplishments achieved by our research group in the past 10 years with respect to the HR-Si interposer applied to 3D heterogeneous RF integration are expanded, including the fundamental research on the HR-Si interposer design and process validation, which aims at addressing the core problem of high-frequency loss, new 3D RF interconnection structure design and process validation, and application and verification with 3D heterogeneous RF integration to demonstrate its feasibility and technical advantages. In order to elevate the integration level, an HR-Si interposer integrated inductor and integrated patch antenna on stacked HR-Si interposers were studied. Given the heat dissipation problem brought by the high-performance TR module, an HR-Si interposer with an embedded microchannel was investigated and validated with a 2–6 GHz GaN HEMT-based power amplifier.

In addition, TGV interposer technology is also reviewed along with research and development advancements in TGV formation, metallization, TGV-enabled IPDs, and TGV interposer-based 2.5D/3D RF integration, to provide a full view for interposer technology for 3D heterogeneous RF integration.

References

[1] Beica R. 3D integration: applications and market trends. In: IEEE international systems integration conference; 2015.
[2] Cognetti C. The impact of semiconductor packaging technologies on system integration an overview. In: European solid-state circuits conference; 2009. p. 23–7.
[3] Zhou X. Review of advanced packaging. Integr Circ Appl 2018;35(297(6)):8–14 [in Chinese].
[4] Tang X. Development and application of microsystem technology. Modern Radar 2016;12 [in Chinese].
[5] Cui K, Wang C, Hu Y. Development and application of 2.5D/3D packaging technology for radio frequency microsystems. Electron Mech Eng 2016;6 [in Chinese].
[6] Liu Y. High resistance silicon IPD technology in system integration. Modern Electron Technol 2014;14:136–9 [in Chinese].
[7] Liu P, Wang J, Lai F. Development trend of military radio frequency integrated circuit technology. Microelectronics 2018;048(005):663–6 [in Chinese].
[8] Li Y. Research on the electrothermal characteristics and optimal design of three-dimensional integrated packaging. Zhejiang University; 2018 [in Chinese].
[9] Yu Y, Zhang H, Huang M, et al. Three-dimensional heterogeneous integration technology of silicon-based RF microsystems. Res Dev Solid State Electron 2019;3 [in Chinese].
[10] Samanta K, Pushing K. The envelope for heterogeneity: multilayer and 3-D heterogeneous integrations for next generation millimeter-and submillimeter-wave circuits and systems. IEEE Microw Mag 2017;28–43.
[11] Green DS, Dohrman CL, Demmin J, et al. A revolution on the horizon from DARPA: heterogeneous integration for revolutionary microwave\/millimeter-wave circuits at DARPA: progress and future directions. IEEE Microw Mag 2017;44–59.
[12] Malta D, et al. TSV-last, heterogeneous 3D integration of a SiGe BiCMOS beamformer and patch antenna for a W-band phased array radar. In: IEEE 66th electronic components and technology Conference (ECTC), Las Vegas, NV, 2016; 2016. p. 1457–64.
[13] Banijamali B, Ramalingam S, Nagarajan K, et al. Advanced reliability study of TSV interposers and interconnects for the 28 nm technology FPGA. In: Electronic components and technology conference; 2011. p. 285–90.
[14] Ho SW, Yoon SW, Zhou Q, et al. High RF performance TSV silicon carrier for high frequency application. In: Electronic components and technology conference; 2008. p. 1946–52.
[15] Ebefors T, Fredlund J, Perttu D, et al. The development and evaluation of RF TSV for 3D IPD applications. In: IEEE international systems integration conference; 2013. p. 1–8.

[16] Bouayadi OE, Dussopt L, Lamy Y, et al. Silicon interposer: a versatile platform toward full-3D integration of wireless systems at millimeter-wave frequencies. In: Electronic components and technology conference; 2015. p. 973–80.
[17] Mengcheng W, Shenglin M, Han C, et al. Design, fabrication and test of dual redundant TSV interconnection for millimeter wave applications. In: International conference on electronic packaging technology; 2019.
[18] Lamy Y, Dussopt L, Bouayadi OE, et al. A compact 3D silicon interposer package with integrated antenna for 60GHz wireless applications. In: IEEE international 3D systems integration conference; 2013. p. 1–6.
[19] Pares G, Jeanphilippe M, Edouard D, et al. Highly compact RF transceiver module using high resistive silicon interposer with embedded inductors and heterogeneous dies integration. In: Electronic components and technology conference; 2019.
[20] Wang H, Lv Y, Gao W, et al. A study on the characterization of quasi-three-dimensional PN junction capacitor. In: International conference on electronic packaging technology & high density packaging. Beijing, China: IEEE; 2009.
[21] Liu H, Fang R, Miao M, et al. Detection of void density in the through silicon via using artificial neural network. In: International symposium on electromagnetic compatibility; 2019.
[22] Ma S, Chai Y, Yan J, et al. A 2.5D integrated L band receiver based on high resistivity Si interposer. In: International conference on integrated circuits, technologies and applications (ICTA); 2018. p. 74–7.
[23] Shin KR, Arendell J, Eilert K. Compact 5G n77 band pass filter with through silicon via (TSV) IPD technology. In: IEEE 20th wireless and microwave technology conference (WAMICON), Cocoa Beach, FL, USA; 2019. p. 2019.
[24] Takahashi S, Tatsukoshi K, Ono M, et al. TGV technology for glass interposer. In: 2012 4th electronic system-integration technology conference; 2012. p. 1–3.
[25] Töpper M, et al. 3-D thin film interposer based on TGV (Through Glass Vias): an alternative to Si-interposer. In: 2010 proceedings 60th electronic components and technology conference (ECTC); 2010. p. 66–73.
[26] Lee CK, et al. Investigation of the process for glass interposer. In: 2013 8th international microsystems, packaging, assembly and circuits technology conference (IMPACT); 2013. p. 194–7.
[27] Liu Z, et al. Electroless and electrolytic copper plating of glass interposer combined with metal oxide adhesion layer for manufacturing 3D RF devices. In: 2016 IEEE 66th electronic components and technology conference (ECTC); 2016. p. 62–7.
[28] Onitake S, Inoue K, Takayama M, et al. TGV (thru-glass via) metallization by direct Cu plating on glass. In: 2016 IEEE 66th electronic components and technology conference (ECTC); 2016. p. 1316–21. https://doi.org/10.1109/ECTC.2016.231.
[29] Wu Z, Min J, Kim M, et al. Design and demonstration of ultra-thin glass 3D IPD diplexers. In: 2016 IEEE 66th electronic components and technology conference (ECTC); 2016. p. 2348–52.
[30] Min J, Wu Z, Pulugurtha MR, et al. Modeling design fabrication and demonstration of RF front-end module with ultra-thin glass substrate for LTE applications. In: Electronic components & technology conference. IEEE; 2016.

CHAPTER 2

Design, process, and electrical verification of HR-Si interposer for 3D heterogeneous RF integration

2.1 Introduction

Typical TSV interposers for 3D heterogeneous RF integration can be made on HR silicon (HR-Si) substrates, which vary from 1 to $10\,k\Omega\cdot cm$ in resistivity. The geometry of TSV can be cylindrical through hole, irregularly shaped through hole, and so on. The in-hole metallization can usually be achieved by copper (Cu) plating, gold (Au) plating, or tungsten (W) CVD, to manufacture hollow/fully filled Cu interconnection, or W interconnection. Several examples can be shown.

Since 2013, CEA-Leti (France) has reported a sequence of research results on an RDL-first/via-last technique for fabricating hollow TSV interposers on HR-Si substrates. Fig. 2.1A shows the process flow of fabrication [1]. First, an HR-Si wafer is provided whose surface has been thermally oxidized; 10 μm of Cu is electroplated on the oxide as a wiring layer with a pre-deposited seed layer. Coat the metal with a thin film of SiN using CVD, followed by 11 μm of Asahi Glass AL-X polymer to create an organic front side passivation. Open windows on the laminated passivation layer by photography and make bonding pads of Ti/Ni/Au. Then the wafer is temporarily bonded on a glass carrier with ZoneBond glue (Brewer Science) and is thinned down to 180 μm by grinding. A deep reactive ion etching (DRIE) Bosch process is carried out to open TSV holes, which are down to the front side bonding pads. Deposit a conformal SiON layer to achieve insulation between the TSV and substrate. Remove SiON on the bottom of the holes (i.e., back side surface of metal pad) by RIE and electroplate 7 μm of Cu with a predeposited seed layer. Finally, make a passivation layer and bonding pads on the back side, just like what has been done on the front side. In this design the laminated structures on both sides are the same, such that a low warpage of the final substrate can be ensured. The hollow Cu TSV interconnection has a diameter of 30 μm and a depth of 180 μm. The conformal metal layer inside the hole is about 7 μm in thickness. An SEM image of the cross section is shown in Fig. 2.1B. The insertion loss is measured at less than 1 dB at 0-60 with 60 GHz.

In 2013, Silex (Switzerland) proposed an approach to fabricate TSV interposers on HR-Si wafers of 305 μm (resistivity $>3\,k\Omega\cdot cm$), by performing DRIE as well as Cu

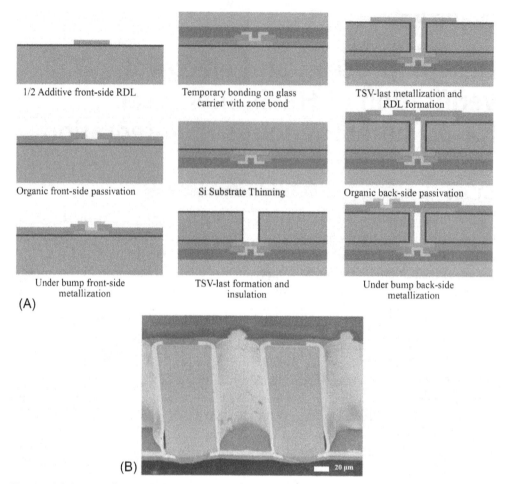

Fig. 2.1 (A) Process flow and (B) cross-section SEM image of the RDL-first/via-last technique route based HR-Si TSV interposer by CEA-Leti (France).

pattern plating on both sides. The process flow is demonstrated in Fig. 2.2A [2]. As shown, the X-shaped TSV consists of two coaxial vias in the form of in-series connections. The larger via is cylindrical with an aspect ratio (AR) of 3:1 ($h = 180\,\mu m \pm 10\,\mu m$; $\phi = 60\,\mu m$), while the smaller one is tapered ($h = 25\,\mu m \pm 10\,\mu m$; opening $\phi = 25\,\mu m$; bottom $\phi = 8\,\mu m$). TSVs of this design had been tested as an RF TSV interconnection connected by coplanar waveguide (CPW). The insertion loss of a single TSV is measured at less than 0.04 dB at 5 GHz.

In order to enhance the high-frequency (HF) electrical characteristics, reliability, and compatibility with the integration of RF microelectronic chips, and considering the high interconnection density of RF integration, Fraunhofer IZM (German), Technische Universität Berlin (German), CIP Technologies (UK), and others reported research results on fabricating hollow Au TSV interconnection on HR-Si substrates in 2016

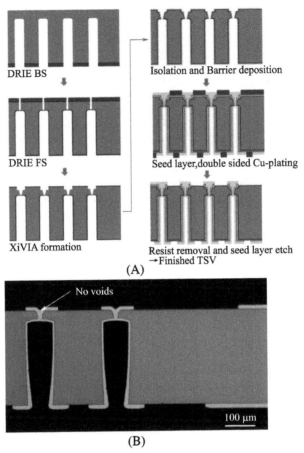

Fig. 2.2 (A) Process flow and (B) example image of the metallized TSV by Silex (Switzerland).

[3]. First, a layer of W/Ti/Ni was sputtered onto the sidewall of vias, which worked as a diffusion barrier between Au and Si. Then the metallization was done by conformal Au electroplating from both sides. An example is shown in Fig. 2.3. The TSV was 200 μm in diameter and 400 μm in depth. A TSV grounded CPW line was designed and tested, and the result showed it had an insertion loss less than 1 dB at 10 GHz. The Au TSV interconnection performed well in reliability, electrical characteristics, and compatibility with the integration of RF microelectronic chips. However, the problem of high cost is significant, while the technique of high aspect ratio Au electroplating is still not matured.

In 2020, CETC 55 (China) reported a new technique for a via-first/RDL-last HR-Si TSV interposer. The conceptual draft and SEM image of part of the interconnection are shown in Fig. 2.4. In general, the interposer is fabricated as follows. Open blind vias on an 8-in. HR-Si wafer (single side polished, resistivity = 10 kΩ·cm). Fill the vias by Cu electroplating, followed by CMP planarization. The TSV should be 20 μm in diameter and 85 μm in depth. Perform a damascene process to deposit 0.6 μm of Cu as grounding layer,

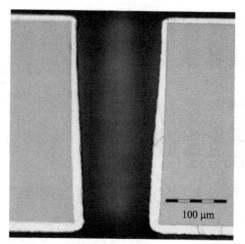

Fig. 2.3 Example image of hollow Au TSV established on HR-Si substrate, by Fraunhofer IZM (Germany), Technische Universität Berlin (Germany), CIP Technologies (UK), and others (2016).

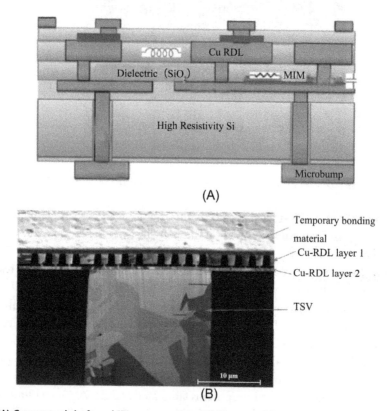

Fig. 2.4 (A) Conceptual draft and (B) cross-section SEM image of the TSV interconnection by CETC 55 (China).

bottom metal layer of MIM capacitor, and bottom connection to multiturn inductor. Establish a sandwich of 0.50 μm TaN/0.60 μm SiN/0.50 μm TaN as a MIM capacitor. Establish a Cu via array (0.5 μm in diameter and 0.5 μm in depth for single via) to connect the bottom metal layer, MIM capacitor, and upper metal layer. Deposit a Cu structure of 2 μm as transmission line and multiturn inductor. Finally, flip the wafer by temporarily bonding to a carrier and perform thinning until the TSVs are revealed.

Experimental tests were made and showed the total insertion loss of the TSV and Cu contact was less than 1 dB at 40 GHz. This new technique route is actually similar to the TSV interposer fabrication process for traditional digital ICs, though HR-Si (rather than low-resistivity silicon) substrate is used and a thick Cu damascene process is introduced to establish the rewiring layer. It is also CMOS-compatible and has a high density in terms of TSV interconnection. However, considering the risk of thermal mechanical failure due to CTE mismatch, the filled Cu TSV interconnection is usually designed 10–20 μm in diameter and 10 in aspect ratio. While the thickness of the TSV interposer is limited to less than 200 μm for most of the cases, the limitation should be introduced as well to achieve its application in 3D heterogeneous RF integration.

The aforementioned RDL-first/via-last technique by CAE-Leti and the via-first/RDL-last technique by CETC 55 are both CMOS-compatible, although some technical challenges should be taken into consideration. As for the former, it is challenging to reveal the metal pad on the front side by etching from the back side, as well as the sidewall passivation of blind vias. As for the latter, the difficulty lies in revealing the filled Cu TSV interconnection. The electroplating and two-sided metallization of open vias, either annular or irregularly shaped, are naturally compatible with the high-density wiring on PCB, and therefore can reduce the cost of equipment. Also, with the technique of open vias it is easier to manufacture an HR-Si interposer of variety of thickness or even with a cavity, such that it can better respond to the demand in the field of 3D heterogeneous RF integration.

In the next part of this chapter, some work done by our research group will be proposed, which includes the design and process of a coaxial ladder-hollow-through TSV interposer, as well as the design and high-frequency electrical characteristics analysis of an RF transmission unit.

2.2 Design and fabrication process of HR-Si TSV interposer

The resistivity of the Si substrate is one of the significant factors related to the HF insertion loss of TSVs. In this experiment, HR-Si wafers with resistivity greater than 2 kΩ·cm were used. Design of a coaxial ladder TSV structure consisting of different-sized hollow vias was employed (Fig. 2.5). During the fabrication process, thermal oxidization was performed to produce SiO_2 as the insulation between TSV metal and the Si substrate. The vias were metallized by conformal Cu electroplating for the transmission of electrical

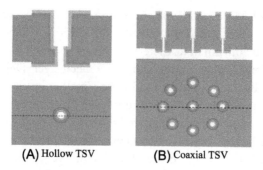

Fig. 2.5 Design of TSV structure in our experiment.

signals. There was also a Ti/Cu barrier/adhesion layer between the plated Cu and SiO$_2$, which avoids the diffusion of Cu into the substrate and increases the adhesion of the Cu with the oxide [4]. Cu was used to establish the electrical connection in a rewiring layer on the surface. Electroless nickel electroless palladium immersion gold (ENEPIG) was used for the bonding pads in order to fit the RF/MW microelectronics integrated devices, while Au contacts are widely used in these chips made of group III-V semiconductor materials. The material parameters are summarized in Table 2.1.

To analyze the thermal performance of the TSV interconnection, a planar stress model (Fig. 2.6) based on linear thermal stress theory is considered, where we neglect effects in the depth direction. The outer and inner radii of the Cu tube are called a and c, respectively, while the outer radius of the Si substrate is called b. Polar coordinates (r,θ) are used for convenience. If the sample is under a uniform temperature load ΔT and the outer boundary is free, as Fig. 2.6 shows, both of the materials are performing thermal deformation. Because Cu has a larger CTE than Si, the deformation amount of the two will be different, such that a thermal stress will be introduced. The trend of deformation and constraint of Si are demonstrated in Fig. 2.6B and C, respectively. The trends of deformation and constraint of Cu are demonstrated in Fig. 2.6D and E, respectively.

Table 2.1 Material parameters of the TSV-based transmission line test structure.

Parameter		Value
HR-Si substrate	Dielectric constant	11.9 F/m
	Thickness	300 μm
SiO$_2$ insulation layer	Dielectric constant	3.9 F/m
	Thickness	2 μm
Cu interconnection	Dielectric constant	1 F/m
	Thickness	8 μm
Au RDL	Thickness	50 nm

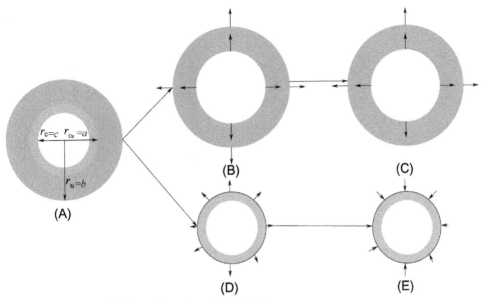

Fig. 2.6 Radial thermal deformation of annular TSV interposer.

Setting $q=b/a$ and $k=c/a$, the components of stress tensor inside the Cu and Si substrate can be calculated by Eqs. (2.1) and (2.2):

$$\left.\begin{aligned}
\delta_r^{Cu} &= \left[1-\frac{k^2}{(r/a)^2}\right]\frac{1}{k^2-1} \times \frac{E^{Cu}E^{Si}(\alpha^{Cu}-\alpha^{Si})\Delta T}{E^{Si}\left(\frac{1+k^2}{1-k^2}-\nu^{Cu}\right)+E^{Cu}\left(\frac{q^2+1}{q^2-1}+\nu^{Si}\right)} \\
\tau_{r\theta}^{Si} &= 0 \\
\delta_\theta^{Cu} &= \left[1+\frac{k^2}{(r/a)^2}\right]\frac{1}{k^2-1} \times \frac{E^{Cu}E^{Si}(\alpha^{Cu}-\alpha^{Si})\Delta T}{E^{Si}\left(\frac{1+k^2}{1-k^2}-\nu^{Cu}\right)+E^{Cu}\left(\frac{q^2+1}{q^2-1}+\nu^{Si}\right)}
\end{aligned}\right\} \quad (2.1)$$

$$\left.\begin{aligned}
\delta_r^{Si} &= \left[1-\frac{q^2}{(r/a)^2}\right]\frac{1}{q^2-1} \times \frac{E^{Cu}E^{Si}(\alpha^{Cu}-\alpha^{Si})\Delta T}{E^{Si}\left(\frac{1+k^2}{1-k^2}-\nu^{Cu}\right)+E^{Cu}\left(\frac{q^2+1}{q^2-1}+\nu^{Si}\right)} \\
\tau_{r\theta}^{Si} &= 0 \\
\delta_\theta^{Si} &= \left[1+\frac{q^2}{(r/a)^2}\right]\frac{1}{q^2-1} \times \frac{E^{Cu}E^{Si}(\alpha^{Cu}-\alpha^{Si})\Delta T}{E^{Si}\left(\frac{1+k^2}{1-k^2}-\nu^{Cu}\right)+E^{Cu}\left(\frac{q^2+1}{q^2-1}+\nu^{Si}\right)}
\end{aligned}\right\} \quad (2.2)$$

Substituting in corresponding material parameters of Cu and Si and simplifying, the equations turn into the form of Eqs. (2.3) and (2.4):

$$\left.\begin{aligned}\delta_r^{Si} &= \left[1 - \frac{q^2}{(r/a)^2}\right]\frac{1}{q^2-1} \times \frac{13.7\Delta T}{12.9 \times \frac{1+k^2}{1-k^2} + 7 \times \frac{q^2+1}{q^2-1} - 2.1} \\ \tau_{r\theta}^{Si} &= 0 \\ \delta_\theta^{Si} &= \left[1 + \frac{q^2}{(r/a)^2}\right]\frac{1}{q^2-1} \times \frac{13.7\Delta T}{12.9 \times \frac{1+k^2}{1-k^2} + 7 \times \frac{q^2+1}{q^2-1} - 2.1}\end{aligned}\right\} \quad (2.3)$$

$$\left.\begin{aligned}\delta_r^{Cu} &= \left[1 - \frac{k^2}{(r/a)^2}\right]\frac{1}{k^2-1} \times \frac{13.7\Delta T}{12.9 \times \frac{1+k^2}{1-k^2} + 7 \times \frac{q^2+1}{q^2-1} - 2.1} \\ \tau_{r\theta}^{Cu} &= 0 \\ \delta_\theta^{Cu} &= \left[1 + \frac{k^2}{(r/a)^2}\right]\frac{1}{k^2-1} \times \frac{13.7\Delta T}{12.9 \times \frac{1+k^2}{1-k^2} + 7 \times \frac{q^2+1}{q^2-1} - 2.1}\end{aligned}\right\} \quad (2.4)$$

Assuming the outer radius of the Si substrate is 120 μm and the inner and outer radii of the Cu tube are 36 μm and 40 μm, respectively, and a temperature difference of 100°C is loaded, the distribution of stress components along the radial direction are calculated and shown in Fig. 2.7. As shown, the maximum stress occurs at the common boundary of Cu and Si. As for the Cu, the changing of both radial and annular stress are steady and nearly linear. As for the Si substrate, inside 60 μm, both radial and annular stress change dramatically, while outside 60 μm, both of them change less significantly and tend to be stable. Fig. 2.8 shows the change of stress at the common boundary with q. It can be

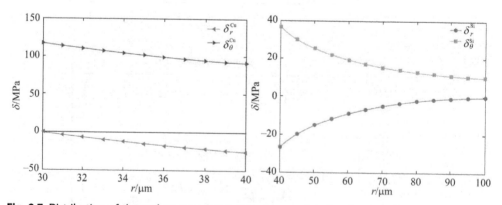

Fig. 2.7 Distribution of thermal stress components along radial direction in hollow annular Cu TSV interconnection, under a temperature difference of 100°C.

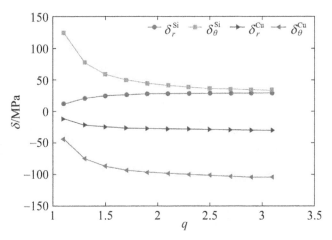

Fig. 2.8 Change of stress at the common boundary of hollow Cu TSV and Si substrate with q.

found that the stress components tend to be stable when the outer radius of the Si substrate reaches twice the outer radius of hollow Cu TSV. In this scenario, adjacent TSVs should be spaced by twice their outer radius to reduce or avoid the coupling of thermal stress. As shown in Fig. 2.9, the value of each of the thermal stress components increases as k decreases. As the Cu sidewall gets thicker, the thermal stress in the radial direction becomes greater; in other words, hollow annular Cu TSV should have higher thermal mechanical reliability than filled Cu TSV at a given condition. The optimization can be determined by using a 3D model. Considering the effects in the depth direction, by insulation layers, and so on, build up a 3D model and perform FEA. Fig. 2.10 shows the thermal stress distribution cloud graph of cylindrical TSV and ladder TSV under a 1 K temperature difference. It can be seen that the

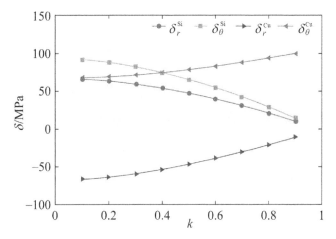

Fig. 2.9 Change of stress at the common boundary of hollow Cu TSV and Si substrate with k.

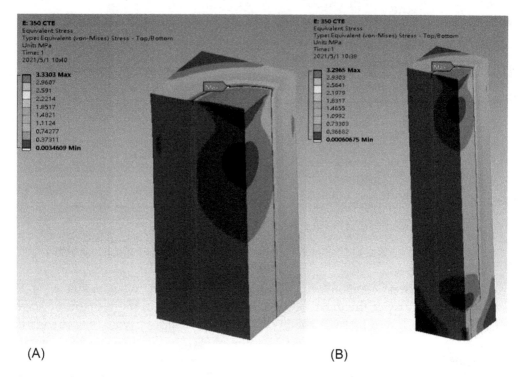

Fig. 2.10 Thermal stress distribution cloud graph of (A) cylindrical TSV and (B) ladder TSV, under a temperature difference of 1 K.

maximum stress in the ladder structure is smaller than that in the cylindrical structure. That implies that the design here is not only beneficial to process, but also performs better in term of thermal mechanical reliability.

Based on the discussed ladder structure TSV, a process for the RF HR-Si interposer is designed (Fig. 2.11). The main technical parameters are summarized in Table 2.2. The general process flow of fabrication is as follows:

(1) An Si wafer of 300 μm, which is double-sided polished and about 3 kΩ·cm in resistivity, is provided.
(2) Etch from both sides to open through vias. Perform DRIE from one and then the other side after photolithography to yield ladder structure TSVs. The large vias should be 80 μm in diameter and 240 μm in depth, while the small vias should be 40 μm in diameter and 60 μm in depth.
(3) Perform RCA standard cleaning to remove any impurity residual, which may contaminate the HR-Si substrate in the following high-temperature process and increase the insertion loss.
(4) Establish high-quality insulation layer. By performing high-temperature thermal oxidation, forming a layer of 100-nm firm SiO_2 on the surface of the substrate as well as the sidewall of vias.

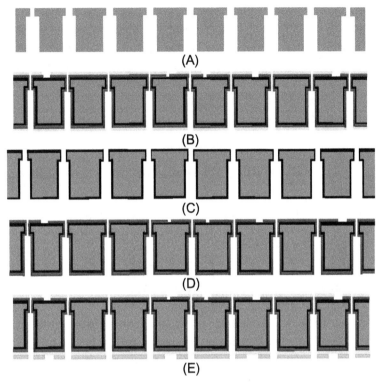

Fig. 2.11 Designed process of HR-Si interposer.

(5) Perform TSV metallization and the establishment of surface rewiring layer. Sputter 200 nm of Ti as adhesion layer and 2 μm of Cu as seed layer onto both sides. Perform photolithography to produce patterned masks. Perform plasma cleaning to remove residue and activate the Cu seed layer. Electroplate 8–10 μm of Cu by controlling the time of performing, concentration of electroplate liquid, current density, etc.

Table 2.2 Main technical parameters for HR Si interposer.

Major parameter	Value
Resistivity of substrate	⩾2 kΩ·cm
Dimension of ladder TSV	ϕ 20–40 μm; AR = 1.5
	ϕ 60–80 μm; AR = 3
Thickness of SiO_2 (TSV sidewall)	0.1 μm ± 0.05 μm
Thickness of SiO_2 (substrate surface)	0.1 μm ± 0.05 μm
Thickness of Cu (TSV sidewall)	5–10 μm
Thickness of Cu (RDL)	8–12 μm
Minimum line width/minimum line spacing (RDL)	70 μm/50 μm
Thickness of Au on Cu RDL	0.05 μm
Thickness of BCB	5–8 μm

Cu TSVs and RDL rewiring lines are formed. Then remove the masks and wet-etch to remove the Cu seed layer and then the Ti adhesion layer.
(6) Plate Ni and Au. Perform electroless Ni plating and Au electroless plating to deposit 3 μm of Ni and 50 nm of Au onto the surface Cu layers on both sides simultaneously.
(7) Passivation. A layer of BCB or dry photosensitive film is coated and patterned on the side of large vias for further bumping.
(8) Screen, dice, and single out.

2.3 Design and analysis of RF transmission structure built on HR-Si TSV interposer

Coplanar waveguide (CPW) is a kind of planar microwave transmission line. Traditionally, it consists of a metal strip at the center and two semiinfinite ground planes parallel at both sides. The ground planes are spaced by some distance from the center strip, as shown in Fig. 2.12A. In analysis of the electromagnetic (EM) field, the spacing and the ground can be regarded as equivalent magnetic barriers and equivalent electric barriers, respectively. In the design of CPWs, all of the grounds are supposed to be in the same electric potential. Because CPW is usually made on double-sided substrates, the grounds can be established by vias and the metal lines on the back side [5].

In this experiment, based on the transmission line model, a CPW structure with grounding TSV array is designed.

Using the full-wave simulation HFSS [6,7], the impact of grounding TSVs on the RF performance of CPW transmission line was simulated and analyzed. As shown in Fig. 2.13, two models are studied and the only difference between them is the existence of grounding TSVs. The structure (A) has a grounding TSV array that plays the role of EM shielding [8], but the structure (B) has no grounding TSV. According to the result of FEA shown in Fig. 2.14, it can be concluded that the existence of grounding TSVs significantly reduces the insertion loss of transmission line. At low frequency, both of the structures perform similarly in terms of S21. However, once the frequency reaches

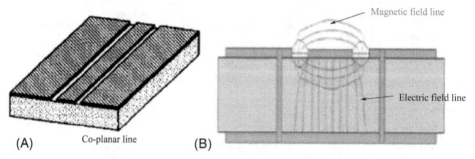

Fig. 2.12 (A) CPW transmission line. (B) Distribution of EM field inside and around CPW.

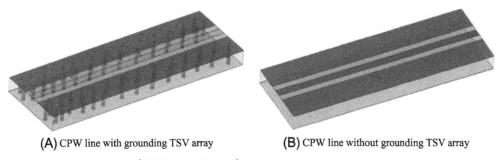

Fig. 2.13 Two structures of CPW transmission line.

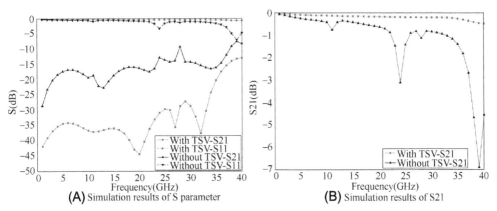

Fig. 2.14 FEA results of two structures.

10 GHz and goes beyond, the loss of the CPW line without grounding TSV becomes obviously larger than that of CPW line with grounding TSV. At 40 GHz, the difference in insertion loss can be couples of dB. Thus the grounding TSV array is able to prevent the drop in insertion loss S21 caused by the occurrence of spurious propagation modes [9], and therefore improve the RF transmission performance. Furthermore, the distribution of electric potential in the grounded plane is more uniform with the existence of grounding TSVs. When grounding TSV is crowded together and close to the CPW signal line, EM energy is concentrated on the signal line such that a better transmission performance can be achieved.

Based on the proposed structure with grounding TSV array, in order to minimize the loss of CPW by utilizing the grounding vias, further simulations are done to study the effect caused by the change in location of ground TSVs. The model used for HFSS simulation is shown in Fig. 2.15. While the dimensions of substrate, transmission line, and grounding TSVs are kept the same, the location and number density of grounding TSVs are altered. In model (A) the spacing between grounding TSVs is larger. The model (B) has smaller spacing between ground TSVs [relative to (A)] while the density of TSVs

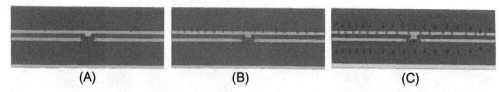

Fig. 2.15 (A) Larger spacing between grounding TSVs, grounding TSVs more distant from signal line. (B) Smaller spacing between grounding TSVs, grounding TSVs less distant from signal line. (C) Density of grounding TSVs doubled.

is higher and the location of the array is closer to the signal line. The model (c) is a revision of (b) that has double the density of grounding vias. The results are summarized in Table 2.3.

As shown, when going from model (a) to (b), the grounding TSVs are getting closer to each other as well as to the signal line, such that the loss decrease. As frequency increases, the amount of loss dropping increases (~1 dB at 40 GHz). Comparing models (b) and (c), the number density of grounding TSVs gets higher, such that a further decrease in transmission loss can be observed. However, such a dropping is not as significant as what was seen between models (a) and (b) when frequency increased (~0.1 dB at 40 GHz).

When the inter-via spacing (i.e., spacing between grounding TSVs) is large, the loss is still relatively large. When the inter-via spacing decreases and/or the number of grounding TSVs increases, the loss decreases. This is because for a larger spacing between grounding vias, there is no strong reflection due to a well-formed electric barrier. In terms of the optimization of via-line spacing (i.e., spacing between grounding TSVs and the signal line), a large spacing will increase the discontinuity in inductance, while a small one will increase the discontinuity in capacitance. As grounding TSVs get closer to the transmission line, the ground-ground resonance at the low-frequency band and high-frequency band (which may impact the high-frequency band) can be well restrained. Also, the grounding TSVs work as the path of the current returning from RF TSVs, such that the loss can be reduced by a shorter path when the via-line spacing decreases [10,11].

Table 2.3 Insertion loss of the CPW model studied at 1–40 GHz.

Frequency (GHz)	S21(dB)			S11(dB)		
	Model(a)	Model(b)	Model(c)	Model(a)	Model(b)	Model(c)
5	−0.1000	−0.0976	−0.0923	−26.1718	−28.4590	−30.8474
10	−0.2091	−0.1485	−0.1405	−17.3488	−30.3639	−36.5434
15	−0.3360	−0.1874	−0.1791	−14.1208	−30.8802	−34.7983
20	−0.5149	−0.2207	−0.2104	−11.5443	−32.5281	−36.3894
25	−0.8091	−0.2487	−0.2322	−9.1318	−29.9368	−33.5911
30	−1.2455	−0.3046	−0.2793	−6.9963	−21.3852	−22.3071
35	−0.8016	−0.3105	−0.2920	−10.5802	−28.5461	−25.2026
40	−1.5197	−0.4139	−0.3605	−6.3272	−18.7523	−19.0388

In other words, at high frequency, the location of the grounding vias will impact the performance in loss significantly. Grounding TSVs should be designed as close to the signal line as possible, while the inter-via spacing should be no more than one-quarter wavelength at the highest operating frequency. In the subsequent part of this experiment, the design is going to follow the scenario of model (c). The inter-via spacing and via-line spacing are set to 200–250 μm and 80–100 μm, respectively.

In order to further evaluate the electrical performance of coaxial ladder-hollow-through TSV interconnection at high frequency, a group of test structures were designed. The test structures are CPW transmission lines with grounding TSV arrays of the same width but different lengths, as shown in Fig. 2.16. The 1-mm-long CPW line is for evaluating the loss per millimeter, while a 5-mm-long CPW line is also integrated to calculate the average loss per millimeter. Comparison in HF electrical performance between the structures can be seen in Fig. 2.17. As shown, the loss per millimeter for the 5-mm-long CPW is similar to that for the 1-mm-long CPW at low frequency (<10 GHz). The former is slightly lower than the latter by an amount on the order of 0.01 dB.

The major process to fabricate TSV interconnection includes etching to open through holes, oxidation, photolithography to form patterned RDL masks, and Cu electroplating to metallize the vias and establish RDL. Due to the limitations in terms of equipment, technique, and so on (such as the resolution of mask aligner, etching selectivity of the etching machine, the temperature of thermal oxidation), unavoidable manufacturing error may occur and influence the transmission performance. Considering the current process capability, the manufacturing process may introduce errors in terms of couples of aspect, such as resistivity of substrate, diameter of TSVs, and thickness of oxide and metal. Using HFSS simulation, such errors and their impacts on the electrical transmission performance of TSVs are analyzed.

Fig. 2.18A shows an HFSS model to study the effects by degradation of substrate resistivity. The substrate is set to be 1.5, 2.0, and 2.5 kΩ·cm in resistivity for individual simulations, respectively. The results plotted in Fig. 2.18B show how the resistivity influences the transmission loss. As all the dimension parameters of the structure stay

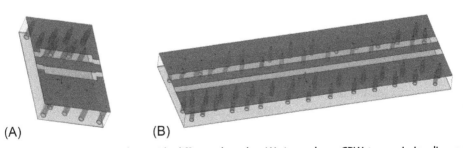

Fig. 2.16 CPW transmission line with different lengths. (A) 1-mm-long CPW transmission line and (B) 5-mm-long CPW transmission line.

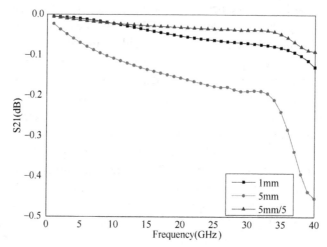

Fig. 2.17 Comparison in loss between the 1-mm-long CPW and 5-mm-long CPW; data acquired by simulation.

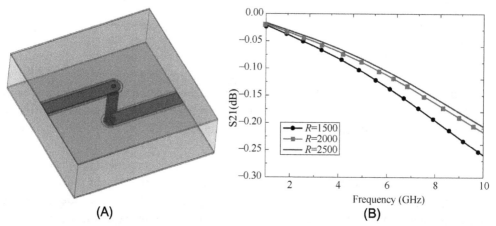

Fig. 2.18 (A) Structure for HFSS simulation. (B) Impact on insertion loss of transmission line caused by the change in substrate resistivity.

the same but resistivity of the substrate decreases from 2.5 to 1.5 kΩ·cm, the insertion loss during transmission increases by 0.07 dB at 10 GHz.

Fig. 2.19A shows the model of a 1-mm-long CPW line with grounding TSV array. The radius of vias is set to be (40 ± 6) μm and the impact on insertion loss caused by such a change was studied. The results plotted in Fig. 2.19B show that the effect is small when the fluctuation in TSV radius is slight. Fig. 2.19C shows that the slight fluctuation in metal thickness can barely impact the insertion loss. However, as shown in Fig. 2.19D, the thickness of the SiO_2 layer does have a nonnegligible impact on insertion

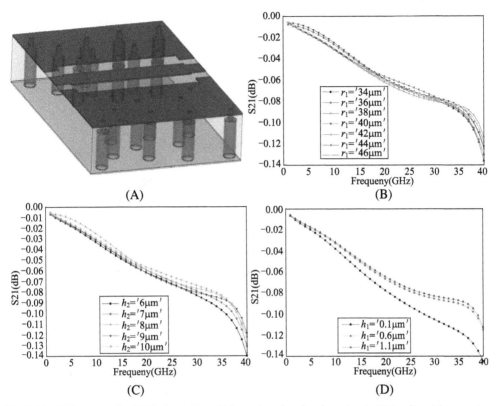

Fig. 2.19 (A) Structure for HFSS simulation. (B) Impact on loss by changing in TSV radius. (C) Impact on loss by changing in Cu thickness. (D) Impact on loss by changing in SiO_2 thickness.

loss. In individual simulations, the thickness of oxide is set to 0.1, 0.6, and 1.1 μm, respectively. While the difference seems to be insignificant at low frequency, as frequency goes up, the effect due to different oxide thickness becomes obvious. The oxide of 0.1 μm yields the largest loss. When the thickness of SiO_2 increases to 0.6 μm, the loss at 40 GHz decreases by 0.03 dB. When the thickness increases to 1.1 μm, a further but slight drop (by 0.003 dB) in loss at 40 GHz can be observed.

2.4 Research on HR-Si TSV interposer fabrication process
2.4.1 Double-sided deep reactive ion etching (DRIE) to open HR-Si TSV

A Si wafer of 300 μm, which is double-sided polished and about 3 kΩ·cm in resistivity, is used. The traditional Bosch process is chosen and the Multiplex ICP Etcher produced by STS (Surface Technology Systems) is employed to perform DRIE. The Bosch process is essentially an anisotropic etching process in a scenario of periodically repeated etching and passivation. SF_6 gas is used for etching while C_4F_8 gas is used for passivation. The

 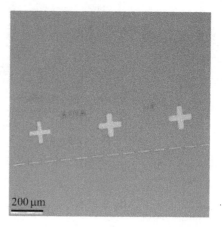

Fig. 2.20 Surface of wafer after DRIE process.

passivation gas will be deposited on the sidewall of TSV to ensure the etching rate in the vertical direction is larger than that in the horizontal direction, such that the etching can be principally performed vertically. Fig. 2.20 shows the surface of the wafer after etching.

2.4.2 Thermal oxidation to form firm insulation layer

Perform RCA standard cleaning after etching, then put the wafer into the oxidation furnace. The temperature for the oxidation process is 700–1400°C. Supply oxygen (O_2) through an oxidation pipeline to drive the surface of the wafer into reaction and form a layer of SiO_2. The oxide layer formed in this way (thermal oxidation) is characterized by the following properties.

(1) The oxide layer is firm and therefore able to prevent more oxygen or water molecules from oxidizing the Si substrate at room temperature.
(2) The oxide layer is attached to the substrate tightly, while it has excellent chemical stability as well as electric insulation.
(3) The oxide layer only reacts with HF but not any other acid. The chemical equation for the reaction of SiO_2 with HF is:

$$SiO_2 + 6HF = H_2(SiF_6) + 2H_2O$$

where $H_2(SiF_6)$ is a water-soluble complex.

The thermal oxidation process can be classified as dry oxidation or wet oxidation. Dry oxidation yields relatively thin oxide layers, while the wet oxidation can yield a thick oxide layer at a high reaction rate. However, several experiments have shown that oxide produced by dry oxidation is more firm and surface-favorable. As the insulation layer plays an important role of electrical isolation inside the RF interposer, the quality of SiO_2 can directly determine the loss during transmission. Therefore in this experiment

Table 2.4 Wafer resistivity monitoring results at different stages of the process.

Stage of process	Wafer resistivity (kΩ·cm)				
	Point 1	Point 2	Point 3	Point 4	Point 5
Original wafer	2.469	2.258	2.158	2.303	1.995
After etching	2.363	2.045	2.096	2.115	1.780
After oxidation and electroplating	2.094	1.840	1.759	1.899	1.658

dry oxidation is chosen for oxide formation. It took 7 h to perform the process. The produced oxide layer was measured at approximately 100 nm using a thickness gauge, which meets the technical demand.

Etching, oxidation, and other processes might introduce damage or contamination into the wafer, which can lead to degradation in wafer resistivity and the RF performance. Hence, it is critical to monitor and control the wafer resistivity throughout the whole fabrication process. A probe station is used to measure the resistivity of the HR-Si wafer. Considering the in-wafer uniformity, a five-point measurement method was adopted. The results of the measurement are recorded and summarized in Table 2.4. As shown, the degradation is essential but not obvious.

2.4.3 Patterned Cu electroplating to achieve metallization and establish RDL layer

In order to coat the surface of the wafer as well as the sidewall of the vias, a physical vapor deposition (PVD) process was adopted. Perform PVD on the large-via side and then the small-via side to deposit 500 nm Ti/2 μm Cu as adhesion/seed layers, such that the reliability of metal sidewall interconnection can be enhanced. Fig. 2.21 shows the interposer after double-sided sputtering.

Perform photolithography to form electroplating masks on both sides of the seed-layer-sputtered wafer. In order to ensure the precision of metal patterning, the expose dose and expose time should be carefully controlled; otherwise the electroplating patterning may be unnecessarily enlarged. It is also important to ensure the continuity of the TSV sidewall (formed by the following electroplating process) by completely removing the residual photoresist inside vias. Several studies in prebake temperature, exposure time, developing time, and so on had been carried out and the process parameter was already determined.

Fig. 2.22 is the microscopic image of the wafer patterned by photoetching. The precision of mask outlines was ensured and none of the pattern is distorted. The photoresist is uniform in thickness and there is no leaking or floating. In-wafer thickness of the photoresist is measured (12 ± 1) μm using a profilometer, which meets the requirement of subsequent Cu electroplating.

Fig. 2.21 X-ray image of the interposer after double-sided sputtering.

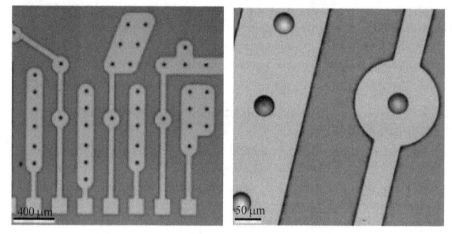

Fig. 2.22 Microscopic image of the wafer patterned by photolithography.

Also, a dry film photoresist is used to make electroplating masks. The dry photoresist chosen is HP-3038, which is (20 ± 2) μm thick by itself. HP-3038 is a photoresist, semiconductor photo area is required when forming the masks. Fig. 2.23 is the microscopic image of the wafer on which dry photoresist is applied.

RDL wiring on the surface as well as the metal interconnection on TSV sidewall will be established by adopting Cu electroplating. Once electricity is applied, positively charged Cu ions will move toward the cathode (wafer), where they will gain electrons, get reduced into metal (atoms), and form a plating layer gradually. The chemical equation is:

$$Cu^{2+} + 2e^- = Cu$$

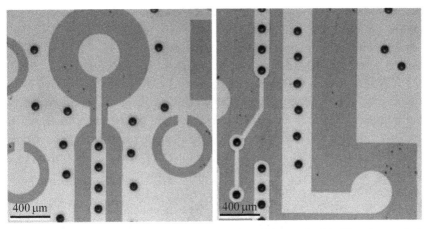

Fig. 2.23 Microscopic image of the wafer patterned by photolithography (dry film photoresist is used).

In this experiment both sides of the wafer will be electroplated simultaneously. The major challenge in such a process is to deposit Cu continuously and conformably to form a hollow metal interconnection on the TSV sidewalls. While the electroplating area on the surface of the wafer is large, the TSVs are small (in diameter) and deep. The distribution of current over the target surface of electroplating is very like to be nonuniform. If the current density is larger at the open end of the vias, what's worse is that the convex end grows faster than the concave end in the high current density area; they are thus probably blocked outside before sufficient Cu ions can reach the TSV sidewalls. Such a failure will lead to discontinuity in metal interconnection. The ingredients of electroplate liquid and the flow field during the process should be precisely controlled in order to avoid this situation.

Apart from the virgin makeup solution (VMS) of acidic copper sulfate, the liquid used in TSV electroplating also contains a variety of additives specifically for through-hole electroplating, such as leveler, suppressor, and accelerator. In general suppressors are polyethylene glycol with different molecular weight (uncharged), accelerators are SPS dithiodipropane sulfonate and so on, which contain sulfur (negatively charged), suppressors are quaternary ammonium salt which contain nitrogen (positively charged). The concentration of ions and additives can significantly impact the quality of electroplating. Moreover, the additives will be consumed, and the consumption rate varies. The concentration of plating additives in the plating process should be carefully monitored and needs to be supplemented on time.

The concentration of Cu ions was measured using a spectrophotometric or chemical titration method. The concentration of sulfate ions was evaluated by the change in color after reacting with a chromogenic reagent. As for the chloride ion, silver nitrate is added into the test solution to trigger a reaction that yielded a cloudy product. Concentration of the cloudy product is proportional to the concentration of chloride ions; therefore the chloride ion concentration can be precisely measured using a scattered light concentration meter despite the influence of chromogenesis.

The concentration of additives can be measured by adopting cyclic voltammetry stripping (CVS), in which a scanning voltage is cyclically altered such that a CV characteristic curve can be obtained. When the working voltage is negative (positive), the working electrode plays the role of cathode (anode). In this scenario, metal will be electroplated onto and stripped from the electrode in a single cycle of voltage alternation. The amount of metal electroplated in a specific time duration can be characterized by the area (Ar) of the peak enclosed by the x-axis and curve of the current during stripping, while this area is set to be Ar^0 for the pure VMS.

The measurements for three kinds of additives are generally based on the relationship between their concentration and the electroplating rate, which can be plotted as curves, as shown in Fig. 2.24. Dilution titration is adopted for the measurement of suppressor concentration, as follows. First, add some suppressor of known concentration into the VMS and measure to obtain a standard curve. Then add the determinand suppressor into another portion of VMS and measure to obtain a new curve. Compare the two curves to evaluate the concentration of suppressor. A modified linear approximation technique (MLAT) is adopted for the measurement of accelerator concentration, while the relationship curve is linear for a low accelerator concentration. First, add an overdose of suppressor to saturation, such that its influence can be neglected. Then add some accelerator of known concentration, followed by some determinand liquid. Finally add some accelerator of known concentration again. From the slope of the relationship curve, accelerator concentration can be evaluated. The measurement for leveler concentration is generally the same as the measurement for suppressor concentration, though the assisting solution containing an overdose of suppressor and accelerator should be prepared first in order to eliminate their impact.

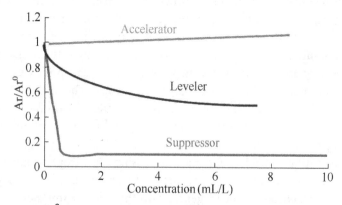

Fig. 2.24 Impact on Ar/Ar^0 due to additive concentration.

Qualilab is used for additive concentration measurement. For the reference electrode, a mixture of $n(10\% H_2SO_4) : n(KCl) = 20:1$ is used as the internal standard liquid, and 10% H_2SO_4 is used as external standard liquid.

Before the experiment, prepare everything that might be used, which includes determinand solution, standard solution (VMS + 12 mL/L suppressor + 10 mL/L leveler + 12 mL/L accelerator), pure VMS, accelerator assisting solution (58 mL of VMS + 2 mL of leveler), leveler assisting solution (56 mL of VMS + 2 mL of suppressor + 2 mL of leveler), pipette, pipette tips, beaker, volumetric flask, measuring cylinder, injector, and so on.

During the experiment the following points should be closely attended to.

(1) Stimulate the electrode before testing each of the additives. At first, place the beaker containing 40 mL of VMS on the operating floor. Use an elevator to raise the beaker and ensure each of the electrodes is immersed in the solution by no less than 0.5 cm. Perform the electrolysis cycle for five times at least. The cycle should break automatically as the stability is higher than 99.6% when Ar is supposed to be (37 ± 1) mC (the fluctuation may be larger than ± 1 mC due to environmental factors like temperature). Do further stimulation by performing electrolysis for 20 times at least, if any result is irregular.

(2) Keep the environmental temperature at $(25 \pm 0.5)°C$.

(3) After each time replacing the solution inside the injector, place the empty beaker on the operating floor. Move the new solution into the beaker by using an injector and replace the liquid inside the conduit of the analyzer.

(4) Before testing the electroplate liquid, stir at 2500 r/min for 15 s to achieve uniformity.

Standard curves are supposed to be plotted before the measurement of suppressor concentration. Move the standard solution prepared into centrifuge tube 1 and perform the replacement. After the electrodes are immersed in 50 mL of VMS, start the measurement and get the first Ar/Ar^0 value. In the same scenario, read out and record an Ar/Ar^0 value every time 0.05 mL of standard solution is automatically added. Stop measuring when the value becomes less than 0.1 and draw the standard curve. The result at the left bottom of the screen shows the calibration factor is 0.0729. The testing for the determinand liquid is no different, except the standard solution is replaced by the determinand. The result of the experiment shows the determinand suppressor has a concentration of 8.17 mL/L (Fig. 2.25).

The standard curve of the leveler is plotted as follows. Start the measurement after immersing the electrode in the leveler assisting solution. Manually add 2 mL of standard electroplate liquid for many times, until the Ar/Ar^0 value becomes less than 0.7. The testing for the determinand liquid is no different, except the standard liquid is replaced by the determinand. Comparing the two curves, the concentration of the determinand leveler is found to be 10.6 mL/L (Fig. 2.26).

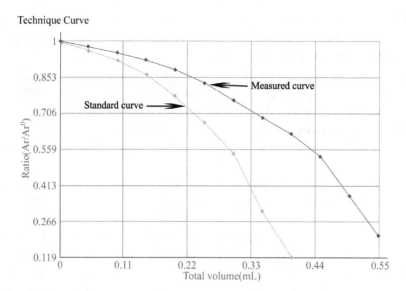

Fig. 2.25 Result of suppressor concentration measurement.

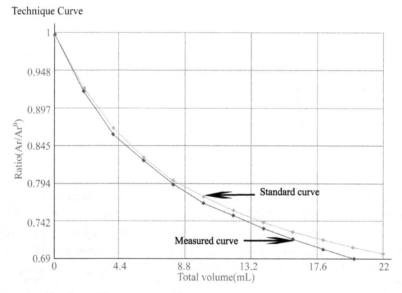

Fig. 2.26 Result of leveler concentration measurement.

There is no need to draw the standard curve for the testing of the brightener. Move the brightener into centrifuge tube 2 and perform the replacement. Place the beaker containing 60 mL of accelerator assisting solution on the operating floor and start the electrolysis cycle. The system will measure the intercept point. Manually add 20 mL of determinand solution and obtain the second point. In subsequence, the system will

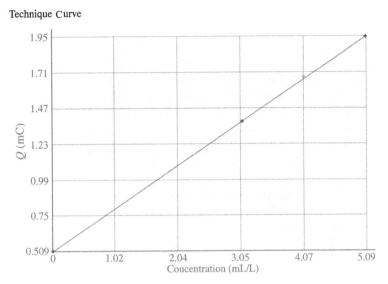

Fig. 2.27 Result of accelerator concentration measurement.

automatically add 0.01 mL of accelerator so that the third and fourth points can be obtained. Check the curve once the experiment is finished. A linearity of 1.0±0.8 should be the correct result. As shown in Fig. 2.27, the concentration of the accelerator is 12.4 mL/L.

Based on the test results, adjust the concentration of the electroplate liquid to standard. Perform plasma cleaning of the surface of the Cu seed layer to remove residual photoresist and improve the wettability. Then the electroplating can be performed by using the equipment shown in Fig. 2.28. After loading the wafer, clean to remove contaminants such as oxide, organics, and dust. Rinse to remove the liquid from the cleaning tank. Preimpregnate to form a microscopically rough surface on the Cu such that the bonding between the electroplating layer and seed layer can be strengthened.

The goal is to electroplate a Cu layer of (10±2) μm. The time and current of electroplating can be figured out by using Eqs. (2.5) and (2.6). A coulomb of charge is equal to 1 A of current passed in 1 s. Faraday's Law: 1 F or 96,500 C is required to deposit 1 g equivalent weight metal. So 1 g equivalent weight metal equals to 1 g atomic weight metal divided by its valence. That means 1 A flowing for 1 s represents 1 C. Faraday's law states that 96,500 C (1 F) will deposit 1 g equivalent weight of a metal. Equivalent weight is the atomic weight of the metal divided by its valence (Eq. 2.7). Two-stage electroplating is adopted. Perform preplating with a small current at first to sufficiently impregnate the surface and vias by liquid. Then apply a large current to thicken the Cu layer quickly. The sprinkler should be turned on and swung during the process, while the sprinkling can facilitate the liquid getting into vias,

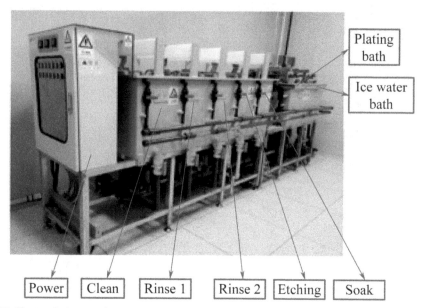

Fig. 2.28 Electroplating equipment.

and the swinging can ensure the wafer contacts with the electroplate liquid sufficiently so that the plating can be more uniform. Take the wafer out and rinse with deionized water once the electroplating is finished. Blow-dry with nitrogen and then bake to dry using vacuum oven. Use a microscope to check if there is any RDL wiring deformed or if there is any photoresist leaked. Then measure the thickness as well as the surface roughness of electroplated Cu using a profilometer. Check the metallized sidewalls of TSVs by adopting X-ray imaging. As shown in Fig. 2.29, continuous metal interconnections are formed. The sample was ground and checked using SEM. As shown in Fig. 2.30, metal layers inside the TSVs are uniform and continuous. The electroplating process reaches the expectation.

$$\begin{aligned}
\text{Thickness of electroplated layer} &= \text{Current density (ASF)} \\
&\quad \times \text{Time of electroplating (min)} \\
&\quad \times \text{rate of electroplating} \times 0.0202, T(\mu m) \\
&= CD(ft^2) \times t(\min) \\
&\quad \times 0.0202 \text{ (the efficiency of electroplating copper} \\
&\quad \text{can generally reach 100\%)}
\end{aligned} \quad (2.5)$$

$$\text{Current} = \text{Current density (ASD)} \times \text{Area}(dm^2), I(A) = CD(A/dm^2) \times S(dm^2)$$
$$(2.6)$$

 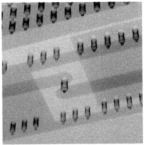

Fig. 2.29 Product of electroplating (X-ray image).

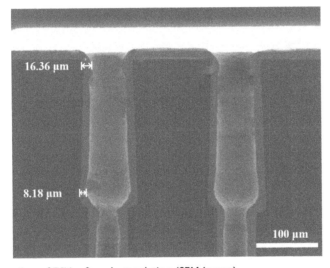

Fig. 2.30 Cross section of TSVs after electroplating (SEM image).

$$\frac{96500 \times V}{M} = \frac{I(A) \times t(s)}{m(g)} \quad (V = \text{valence}, M = \text{molar mass}) \tag{2.7}$$

Remove masks with acetone and alcohol. Spray photoresist onto the wafer to protect the patterned area before etching the seed layer; otherwise the electroplated Cu may also be etched during the process. Since the metal layer inside the TSV should be particularly protected, photoresist should be deposited sufficiently inside the vias. Remove photoresist outside the area of electroplating.

The liquid for Cu etching is a mixture of acetic acid:H_2O_2:water in the ratio of 1:1:20, which is safe and has a low etching rate such that it is easy to control the depth of etching by controlling the etching time. Sometimes additives are also used in the process. The liquid for Ti etching is a mixture of hydrofluoric acid:water in the ratio of 1:50, which also has a low etching rate (for Ti).

2.4.4 Electroless nickel electroless palladium immersion gold (ENEPIG)

The principle of electroless Ni/Au plating is depositing (plating) a layer of Ni by autocatalytic reaction on the surface of a material, and then plating a layer of Au on the Ni by replacement reaction [12,13]. Patterned photoresist masks should be used to cover the part where there is no need to perform the process. The part needing to be plated should be developed and revealed. Several pretreatments should be done before plating: cleaning (to remove any impurity or dust); microetching (with acidic solution, to remove the copper oxide), acid-leaching, stimulation, postimpregnation, and so on. Cleaning should also be performed after Ni plating and immediately followed by Au plating. The electroless plating is also a double-sided process in the experiment (just like the electroplate performed before). The wafer should be checked after the process to see if there are any metal lines incomplete, distorted, or sticking to each other. Fig. 2.31 shows a microscopic image after Au plating and before removing the photoresist.

2.4.5 Surface passivation

A variety of materials can be chosen to make the surface passivation layer, such as photosensitive BCB [14], photosensitive PI, photosensitive dry film, and so on. BCB and PI are liquids, which are usually applied by spin coating or spray coating so that a thin layer (of micrometers) can be produced. However, it is relatively difficult for these liquid materials to enter the TSVs or for protection to be provided for the sidewall metal. New materials like photosensitive dry film can be used to seal the TSVs or fill the vias conformally by performing high-temperature reflow, but they are relatively thicker so the photolithography pattern might not be precise enough. For this experiment, while the window that is supposed to be opened on the passivation layer is only for the subsequent bumping, regular BCB should meet all of the requirements.

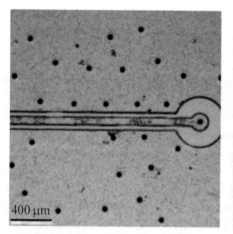 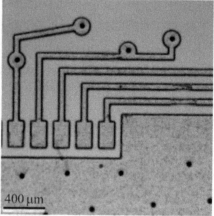

Fig. 2.31 Surface of wafer after electroless Au plating.

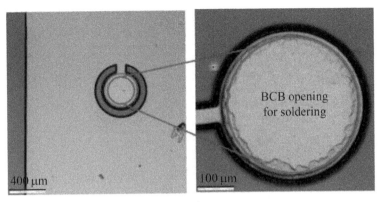

Fig. 2.32 BCB passivation layer (microscopic image).

The process flow of BCB photography is as follows:
(1) Spin coat the AP3000 adhesive at 300 r/min for 40 s.
(2) Spin coat BCB at 3000 r/min for 40 s. The BCB should be removed from the refrigerator 24 h in advance to defrost.
(3) Prebake at 75°C for 90 s.
(4) Expose for 15 s.
(5) Develop using DS3000 at 36°C for 4 min, followed by using DS3000 at room temperature for 2 min. The developing process should be performed in a yellow light room.
(6) Rinse with water. Since the BCB after developing is grease-like and barely soluble in water, a strong water stream should be applied until the grease-like material is removed.
(7) Postbake to solidify BCB. The temperature should be laddered up: from 25°C to 100°C, then keep for 5 min, from 100°C to 150°C, then keep for 5 min, from 150°C to 200°C, then keep for 5 min, from 200°C to 250°C, then keep for 5 min. After keeping at 250°C for another 1 h, the temperature should be decreased to room temperature slowly. Fig. 2.32 is the microscopic image of the wafer after BCB passivation. The quality of the passivation layer is considered good.

2.5 Electrical characteristics analysis of transmission structure on HR-Si TSV interposer

The transmission loss of RDL transmission lines and TSV interconnections is an important indicator to evaluate the RF performance of the interposer. In this research, equipment such as a microwave network analyzer (Agilent N5244APNA-X), RF cables, manual probe station, high-frequency probe (ACP40-A-GSG-150), and so on are used to test the electrical characteristics of the sample, summarized in Table 2.5. The testing

Table 2.5 List of equipment for RF testing.

Equipment	Amount
Microscope	1
Multimeter	1
Vector network analyzer	1
Probe station	1
High-frequency probe	1

circuit is connected as shown in Fig. 2.33A. The equipment shown in Fig. 2.33B is used to examine the proposed design and analysis [15].

Before testing, calibration and leveling of the measurement system should be done to reduce error.

(1) Leveling should be done at first. Place the leveling plate on the operating floor. Stab the plate with probes. The leveling is achieved if all the three indentations are the same.

(2) Then calibrate the system by using SOLT [16], which includes four types of standards: short, open, load, and through. First, choose the range of frequency to be measured (1–40 GHz). Click on the calibration wizard in the calibration menu, choose SOLT mode and ignore isolation, choose calibration piece, examine full two-port SOLT successively, where the S/O stands for Short/Open while a zero value in S11 means full reflection; L stands for Load, which requires a characteristic impedance of 50 Ω as well as a load impedance of 50 Ω, and T stands for Through, while a zero value in S21 means full transmission. Basically, the SOLT calibration can sufficiently eliminate system error.

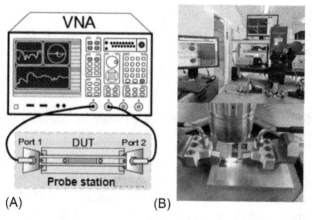

Fig. 2.33 Equipment for RF testing.

Also, several checks should be done for the sample before testing the RF performance.
(1) Check the surface and screen samples using a microscope. Samples with broken or sticking RDL should be excluded.
(2) Check the resistance between transmission line and grounding plane using a multimeter. Such resistance should be large since it should be open between the transmission line and ground. A small (nearly zero) resistance implies that the transmission line and the ground are shorted, and the sample should be excluded.
(3) Check the in-via connection of TSVs using a multimeter. Measure the resistance between the input end and the output end of TSVs. Such resistance should be small (nearly zero) as the interconnection should be continuous. A large resistance implies a discontinuity inside the TSV and the sample should be excluded.

Fig. 2.34A is the microscopic image of the CPW integrated on an HR-Si TSV interposer. The transmission lines are all 5 mm long but have three different line widths. Fig. 2.34B shows the test results for CPW transmission lines with 1 mm length and three different line widths. Fig. 2.34C shows the test results for CPW transmission lines with

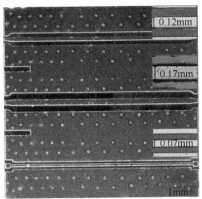

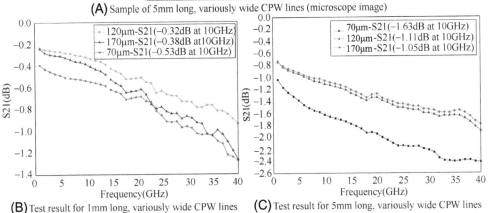

Fig. 2.34 Sample and test result for CPW lines with different widths.

5-mm length and three different line widths. All of the CPWs are grounded by TSV arrays. According to the results, the best-fitting line width for a 1-mm CPW line is 120 μm (VSWR <1.1 at 0.01–40 GHz) while this combination has the smallest insertion loss out of the three. The best-fitting line width for the 5-mm CPW line is 170 μm (VSWR <1.1 at 0.01–40 GHz) while this combination also has the smallest insertion loss out of the three. Fig. 2.35A shows a structure to study the loss of CPW lines with different lengths, by evaluating the average loss per millimeter for 5-mm CPW line and comparing with the loss of 1-mm CPW line, as Fig. 2.35B shows. The test results are summarized in Table 2.6. In general, it can be said that the average loss (per millimeter) for the 5-mm transmission line is less than that for the 1-mm line, which coincides with the computed results. According to the CPW test result, the insertion loss S21 of per millimeter transmission line is less than 0.20 dB/mm at 10 GHz, while an excellent impedance matching can also be observed.

Comparing the simulation result and test result for the 5-mm long, 170-μm wide CPW, an obvious difference can be seen (Fig. 2.36). After reviewing the process parameters, such a difference is tentatively ascribed to the high degree of roughness of the Cu layer in the fabricated sample. Fig. 2.37 shows the test result of surface roughness by using

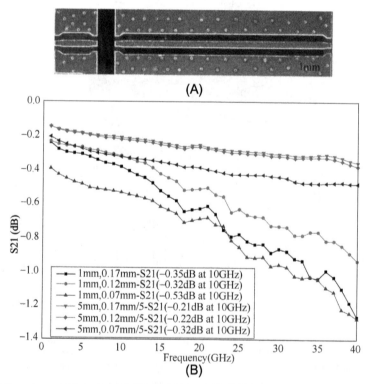

Fig. 2.35 (A) Sample and (B) test result for CPW lines with different lengths.

Table 2.6 Test result of insertion loss per mm for CPW lines with different dimension parameters.

Length	Line width	Average loss 10 GHz	20 GHz	30 GHz	40 GHz
1 mm	0.07 mm	−0.53 dB	−0.69 dB	−0.96 dB	−1.27 dB
	0.12 mm	−0.32 dB	−0.51 dB	−0.73 dB	−0.94 dB
	0.17 mm	−0.38 dB	−0.63 dB	−0.86 dB	−1.26 dB
5 mm/5	0.07 mm	−0.32 dB	−0.39 dB	−0.44 dB	−0.48 dB
	0.12 mm	−0.22 dB	−0.27 dB	−0.32 dB	−0.38 dB
	0.17 mm	−0.21 dB	−0.25 dB	−0.31 dB	−0.35 dB

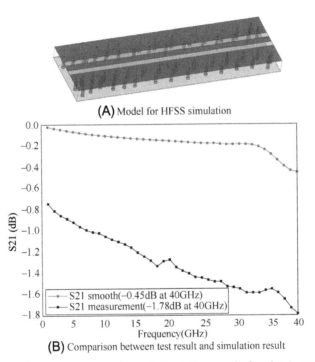

Fig. 2.36 Comparison between test result and simulation result for the insertion loss of CPW transmission line.

a profilometer. Further, HFSS simulation was done by adopting the Huray model. CPW transmission loss in the cases of an ideal (smooth) metal surface and the rough metal surface were studied and compared with the test result (Fig. 2.38B). It can be seen that the metal surface roughness does impact the insertion loss of the transmission line significantly. The loss of the CPW with the rough surface was nearly twice the loss of CPW with smooth surface. The simulation result for the rough surface CPW is relatively closed to the test result. As there are a variety of factors that can impact the performance of insertion loss, detailed studies will be discussed in future chapters.

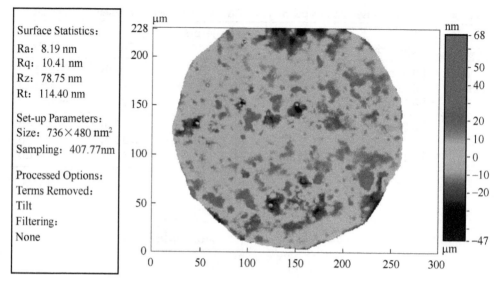

Fig. 2.37 Test result for the surface roughness of Cu layer in sample.

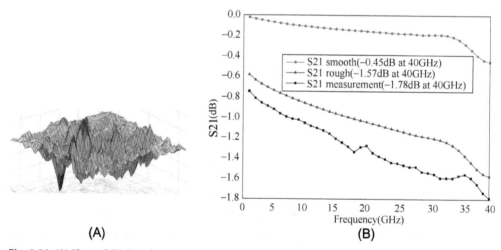

Fig. 2.38 (A) "Spots." (B) Simulation results for rough surface and smooth surface, comparing with test result.

Besides the CPW transmission lines with grounded TSV array, the HR-Si interposer proposed is also compatible with the microstrip. Fig. 2.39 shows the microstrip integrated on the interposer, which is 5 mm long and 250 μm wide. The insertion loss is measured at 1.02 dB at 10 GHz. Fig. 2.40A shows a test structure that consists of a pair of RF TSVs and 6.4-mm-long microstrip, which allows a signal to propagate through the substrate. The test result is shown in Fig. 2.40B. The insertion loss S21 for this structure is −1.3 dB at

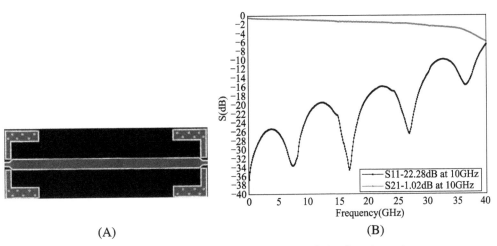

Fig. 2.39 (A) Microscopic image of sample and (B) the test result for the microstrip.

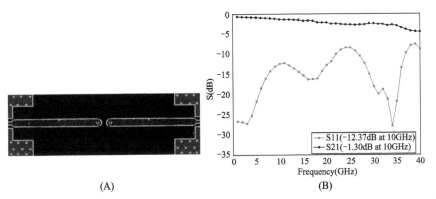

Fig. 2.40 (A) Microscopic image of sample and (B) the test result for the test structure consisting of microstrip transmission line and RF TSVs.

10 GHz. Considering the loss of single RF TSV, S21 per millimeter microstrip at 10 GHz can be estimated to be 0.17 dB.

Table 2.7 summarizes the research results on HR-Si on-board CPW and high-frequency transmission structures around the world. It can be seen that the technique indicators acquired from the test in this research are consistent with other reports from the same period.

2.6 Conclusion

In this chapter, the technique routes for an HR-Si interposer for 3D heterogeneous RF integration are systematically summarized. A ladder structure TSV interconnection consisting of coaxial vias is proposed and its thermal design is analyzed. A CPW transmission

Table 2.7 Test results of CPW lines and TSV interconnection presented in recently published paper.

References	S21 for single TSV or per mm transmission line	S21 for the test structure	Number of transmission vias	Length of line	Characteristic dimension parameter
[17]	0.37 dB at 10 GHz (per TSV)	—	—	—	10 μm/100 μm (Cu-filled TSV)
[18]	—	1 dB at 10 GHz	2	3 mm	150 μm/220 μm [on thin multicrystal silicon wafer or thick (400 μm) polymer substrate]
[19]	0.2 dB at 20 GHz (CPW/mm)	0.3 dB at 25 GHz	2	2 mm	60 μm/200 μm (Cu filled TSV)
[20]	0.05 dB at 10 GHz (per TSV)	0.7 dB at 10 GHz	2	1 mm	100 μm/300 μm (multicrystal silicon, passivation with oxide, Cu filled TSV)
[21]	0.04 dB at 5 GHz (per TSV)	0.182 dB at 5 GHz	2	2.350 mm	90 μm/280 μm (thick, X-shaped TSV with no voids in metal)
[22]	—	0.35 dB at 20 GHz	2	—	75 μm/200 μm (Cu filled TSV)
[23]	—	0.6 dB at 60 GHz	2	—	60 μm/120 μm
[24]	0.07 dB at 20 GHz (per TSV) 0.41 dB at 20 GHz (CPW/mm)	0.55 dB at 20 GHz	2	1 mm	5.5 μm × 13 μm/50 μm (W filled TSV)
[12]	0.53 dB at 75 GHz (per TSV)	—	—	—	AR = 6, magnetic self-assembly and low-k polymer BCB insulation based on Au/Ni plating
This research	0.02 dB at 10 GHz (per TSV) 0.20 dB at 10 GHz (CPW/mm)	1.50 dB at 10 GHz	8	6.4 mm	80 μm/300 μm (hollow Cu TSV)

line with grounded TSV array based on the proposed TSV structure is designed to work at 1–40 GHz. The process of establishing TSVs and the surface RDL by performing double-sided DRIE and double-sided Cu electroplating is introduced. A sample of the product of the process is shown. The test result of the CPW transmission line with grounded TSV array is analyzed. The transmission performance of the coaxial ladder TSV is acquired, which lays the foundation for further research.

References

[1] Lamy Y, Dussopt L, Bouayadi OE, et al. A compact 3D silicon interposer package with integrated antenna for 60GHz wireless applications. In: IEEE international 3D systems integration conference; 2013. p. 1–6.
[2] Ebefors T, Fredlund J, Perttu D, et al. The development and evaluation of RF TSV for 3D IPD applications. In: 2013 IEEE international 3D systems integration conference (3DIC). IEEE; 2013. p. 1–8.
[3] Kröhnert K, Glaw V, et al. Gold TSVs (through silicon Vias) for high-frequency III - V semiconductor applications. In: 2016 IEEE 66th electronic components & technology conference; 2016. p. 82–7.
[4] Sundaram V, Chen Q, et al. Low-cost and low-loss 3D silicon interposer for high bandwidth logic-to-memory interconnections without TSV in the logic IC. Proc Electron Compon Technol Conf 2012;7(2):292–7.
[5] Fan S, Chu L. Microwave technology and microwave circuit. Beijing: China Machine Press; 2009 [in Chinese].
[6] Nakahira K, Tago H, et al. Minimization of the local residual stress in 3D flip chip structures by optimizing the mechanical properties of electroplated materials and the alignment structure of TSVs and fine bumps. ASME J Electron Packag 2012;134(2), 021006.
[7] Bleiker SJ, Fischer AC, et al. High-aspect-ratio through silicon vias for high-frequency application fabricated by magnetic assembly of gold-coated nickel wires. IEEE Trans Compon Packag Technol 2015;5(1):21–7.
[8] Mondal S, Cho S-B, et al. Modeling and crosstalk evaluation of 3-D TSV-based inductor with ground TSV shielding. IEEE Trans Very Large Scale Integr VLSI Syst 2017;25(1):308–18.
[9] Kinoshita T, Kawakami T, et al. Thermal stresses of through silicon vias and Si chips in three dimensional system in package. ASME J Electron Packag 2012;134(2), 020903.
[10] Sun X, Roda Neve C, et al. A simple and efficient RF technique for TSV characterization. In: 2017 IEEE 67th electronic components and technology 64 conference; 2017. p. 1431–6.
[11] Chen Y-W, Zhang M-S, et al. Low cost wafer level packaging of MEMS devices by vertical via-last process. In: 2016 IEEE 66th electronic components and technology conference. 2016. IEEE; 2016. p. 1803–8.
[12] Yuxin D, Dong W, et al. 3-D integration of MEMS and CMOS using electroless plated nickel through-MEMS-Vias. J Microelectromech Syst 2016;25(4):770–9.
[13] Jinghui X, Ding Z, et al. High vacuum and high robustness Al-Ge bonding for wafer level chip scale packaging of MEMS sensors. In: 2017 IEEE 67th Electronic Components and Technology Conference; 2017. p. 956–60.
[14] Ke W, et al. High-frequency characterization of through-silicon-vias with benzocyclobutene liners. IEEE Trans Compon Packag Manuf Technol 2017;7(11):1859–68.
[15] Rack M, Raskin J-P, et al. Fast and accurate modelling of large TSV arrays in 3D-ICs using a 3D circuit model validated against full-wave FEM simulations and RF measurements. In: 2016 IEEE 66th electronic components and technology conference; 2016. p. 966–71.
[16] Wang N, Zhu Y, et al. High-band AlN based RF-MEMS resonator for TSV integration. In: 2017 IEEE 67th electronic components and technology conference; 2017. p. 1868–73.
[17] Kim N, Wu D, et al. Interposer design optimization for high frequency signal transmission in passive and active interposer using through silicon via (TSV). In: Electronic components & technology conference (ECTC), 2016 IEEE 66th. IEEE; 2016.

[18] Vitale WA, et al. Fine pitch 3D-TSV based high frequency components for RF MEMS applications. In: Electronic components & technology conference (ECTC), 2015 IEEE 65th; 2015. p. 585–90.
[19] Chang-Chien P, Zeng X, et al. MMIC compatible wafer-level packaging technology. In: 2007 international conference on indium phosphide and related materials conference proceedings; 2007. p. 14–7.
[20] Hiramatsu S, Mikawa T. Optical design of active interposer for high-speed chip level optical interconnects. J Lightwave Technol 2006;24(2):927–34.
[21] Limansyah I, Wolf MJ, et al. 3D image sensor sip with TSV silicon interposer. In: Electronic Components & Technology Conference; 2009. p. 1430–6.
[22] Fuchs C, Charbonnier J, et al. Process and RF modelling of TSV last approach for 3D RF interposer. In: Interconnect technology conference and 2011 materials for advanced metallization (IITC/MAM), 47. 2011 IEEE International IEEE; 2011. p. 1–3 [10].
[23] Sekhar VN, Toh JS, et al. Wafer level packaging of RF MEMS devices using TSV interposer technology. In: Electronics packaging technology conference (EPTC), 2012 IEEE 14th IEEE; 2012. p. 231–5.
[24] Jung DH, Kim Y, et al. Through silicon via (TSV) defect modeling, measurement, and analysis. IEEE Trans Compon Packag Manuf Technol 2017;7(1):138–52.

CHAPTER 3

Design, verification, and optimization of novel 3D RF TSV based on HR-Si interposer

3.1 Introduction

To address the problem of high-frequency loss, the substrate of the traditional TSV interposer used for a digital IC is replaced with a HR-Si; process optimization is done as well. At present, the substrate resistivity is in the range of 1–10 kΩ·cm for HR-Si. The insertion loss of hollow Cu TSV can reach 0.04 dB at 5 GHz, and the insertion loss of fully filled Cu TSV can reach 0.05 dB at 10 GHz. Table 3.1 covers a variety of performance parameter data, and technical parameters of a HR-Si interposer that have been studied in recent years.

In addition, efforts engaged to explore novel RF transmission structures based on TSV to reduce RF loss, such as a coaxial TSV structure, are presented. In 2012, IBM Microelectronics proposed the design of coaxial TSV [4], which is composed of the internal transmission conductor and the external grounding conductor, with BCB used as the intermediate layer, as shown in Fig. 3.1. The electromagnetic field of signal transmission only exists between the inner conductor and the shell, which greatly reduces the leakage of electromagnetic signals. At the same time, due to the shielding effect of the external grounding conductor, the coupling noise and external electromagnetic interference are effectively reduced. However, the excellent high-frequency performance of the signal is only demonstrated through simulation analysis in this paper; no manufactured samples or test results were disclosed when published.

Compared with the common TSV structure, the manufacturing process of the coaxial TSV is complicated, and some researchers have proposed an optimized design to overcome this problem. In 2020, the ICT Device & Packaging Research Center proposed the Q-COV coaxial structure [5] as shown in Fig. 3.2. By removing one side of the grounding metal, the length of interconnection during installation was reduced, and the metal bump originally used for connection was eliminated, thus reducing the loss. The actual product is shown in Fig. 3.3. The insertion loss of a single vertical interconnection is less than 0.6 dB at 100 GHz. In 2020, Tsinghua University proposed a coaxial TSV interconnection using silicon dioxide as an insulating layer [6], as shown in Fig. 3.4. The measured results of the coaxial TSV are shown in Fig. 3.5. The measured S21 and S11 are

Table 3.1 Main technical characteristics and technical parameters of HR-Si TSV interposer.

References	Material/structure/critical dimensions	Integrated chips or components	Measured parameters	Target application
CEA-Leti (2013) [1]	Area: 6.5 mm × 6.5 mm; Substrate thickness: 120 μm; TSV diameter: 60 μm, aspect ratio 2:1; UBM pad: Ti/Ni/Au, diameter: 290–325 μm; Solder ball diameter: 80 μm	2 Tx/Rx antennas; 1 RF chip	TSV: S21 < 0.6 dB at 60 GHz; Antenna: −10 dB bandwidth range 57–67 GHz, 65.5 GHz gain 5.55 dBi	60 GHz fast data transmission applications
SELIX (2015) [2]	Silicon substrate thickness: 300 μm; TSV: diameter 90 and 50 μm	Integrated inductor	TSV: S21 < 0.04 dB at 5 GHz; DC resistance: 20 mΩ; Inductance: inductance value up to 5 nH, Q > 30, resonance frequency over 10 GHz	Operating frequency up to 10 GHz, inductors can be applied to radio frequency and power SOC
CEA-Leti (2019) [3]	Silicon substrate thickness: 200 μm; RDL thickness: 7 μm	Four kinds of active CMOS chips (VGA, VCO, modulator and PA); 3D inductor; SMD capacitor	RDL: line width 30 μm, thickness 7 μm; DC resistance: 65 mΩ/mm; TSV DC resistance: 3–4 mΩ	2.5D integrated functional RF-SIP transceiver working at 400 and 900 MHz frequency bands; Inductance measurement: inductance values in the range of 0.5–10 nH and quality factor greater than 20

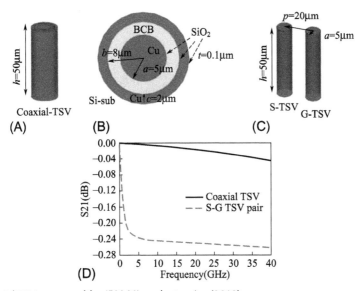

Fig. 3.1 Coaxial TSV proposed by IBM Microelectronics (2012).

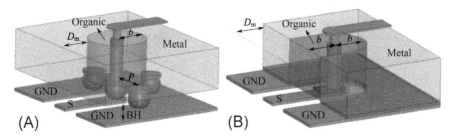

Fig. 3.2 The coaxial interconnection structure proposed by ICT (A) COV and (B) Q-COV (2020).

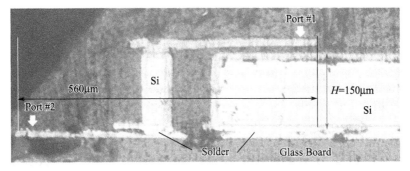

Fig. 3.3 Physical image of Q-COV.

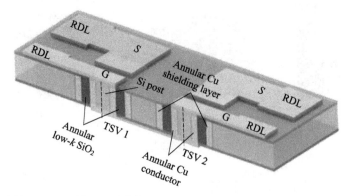

Fig. 3.4 Coaxial TSV proposed by Tsinghua University (2020).

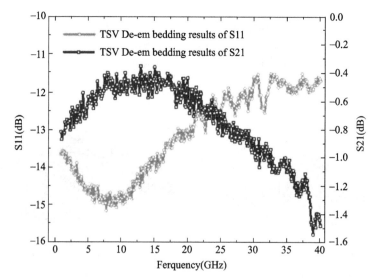

Fig. 3.5 Measured S-parameters of coaxial TSV.

−0.48 and −14.91 dB at 10 GHz, respectively, and −1.48 and −11.73 dB at 40 GHz, respectively. The previously mentioned novel TSV interconnect structures have a low high-frequency loss, but the process is complicated, and is not compatible with the manufacturing process of the normal TSV interposer published in recent years. Moreover, the reliability problem in subsequent practical applications needs to be further studied.

Based on the achievement of the HR-Si interposer described in Chapter 2, this chapter introduces two designs of high-frequency interconnection structure that are compatible with it, including the coaxial TSV structure with low loss and the redundant TSV structure with a high reliability.

3.2 HR-Si TSV-based coaxial-like transmission structure

According to the concept of coaxial TSV, we propose a new vertical coaxial transmission structure, which uses the grounded annular TSV array to replace the external grounding conductor. To verify its feasibility, a test structure of CPW linked by coaxial-like TSVs is designed with impedance of 50 Ω to be capable of working in the range of 0–40 GHz, as shown in Fig. 3.6. The total length of the test structure is 3 mm, the thickness of the substrate is 300 μm, and the line width of the front CPW and the back CPW are 170 and 220 μm, respectively. The design of the outer grounding TSV ring is distanced about 185 μm from the center signal TSV, where all TSV is designed with a radius of 85 μm. In order to facilitate the subsequent test and the extraction of S-parameters of the TSV structure, a CPW line without coaxial-like TSV is designed with a similar total length.

The electrical performance of the coaxial TSV is affected by the radius of the external grounding conductor. HFSS is used to simulate the high-frequency performance of RF TSV with a grounding TSV ring with a changing spacing, where the spacing is called TZR. When the TZR value is set to 0.45, 0.55, 0.65, 0.75, and 0.85 mm, the simulation results are shown in Fig. 3.7. When the frequency is lower than 40 GHz, the insertion losses of several structures are basically less than 1 dB, which is an acceptable level for our application. At the same time, when the TZR is 0.75 mm and 0.85 mm, the insertion loss is relatively small, which may be due to the increase of TSV number in the grounding TSV ring with the distance increases. Crosstalk can change the TSV characteristic impedance and signal propagation speed. Due to the shielding effect of the grounded TSV ring, the electromagnetic field rarely leaks and couples to the adjacent magnetic field. We evaluate the cross-talk immunity of the designed test structure by changing the distance D between two RF TSVs. When the D value is set 0.9, 1.3, and 1.7 mm, the simulation results are as shown in Fig. 3.7. When the distance between the two RF TSVs decreases, the insertion loss does not increase, or even decreases slightly, which indicates that there is less signal interference between the two RF TSVs.

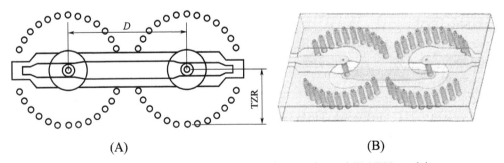

Fig. 3.6 Test structure of CPW+coaxial-like TSV: (A) ichnography and (B) HFSS model.

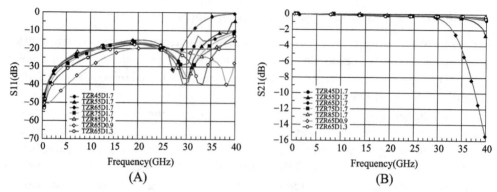

Fig. 3.7 Coaxial-like TSV simulation results: (A) S11 and (B) S21.

3.3 Redundant RF TSV transmission structure

During the manufacturing process for the TSV interposer, defects or failure can occur, including discontinuities or voids in metal holes due to poor sputtering seed layers or electroplating failures, pin-holes, and cracks in TSV oxidation caused by impurities in insulating materials or deposition methods [7]. Discontinuity of the metal will cause the signal channel to disconnect or reflect most of the transmitted signal. Pin-holes in the insulator surrounding the TSV can cause a large leakage current between the TSV and the substrate, further resulting in a resistive short circuit. Voids will cause changes in the interconnection resistance, which results in an increase in signal loss [8]. These defects are decisive for the electrical or RF properties. To reduce the failure risk or damage, the design of redundant TSV has been proposed. The current research on redundant TSV is mainly focused on the 3D logic IC level [9–11]; factoring in that the number of RF TSVs on the package substrate is relatively smaller than 2.5D/3D logic ICs [12,13], there is room for design of redundant RF TSV. However, unlike redundant TSV in logical 3D ICs, the design of redundant RF TSV needs to consider impedance matching. If a defect occurs in an RF TSV, it may bring about a performance change in the entire circuit. In other words, impedance matching must be designed at an acceptable level whatever failure occurs, and this is a problem. Fig. 3.8 shows the designed CPW linked with RF-redundant TSVs and Table 3.2 summarizes the details. Fig. 3.8B and C contain dual redundant TSVs and quad redundant TSVs, respectively. Fig. 3.8A is a reference design with single RF TSV.

Fig. 3.9 shows the S-parameter simulation results of the test structure. When the frequency is 40 GHz, the insertion losses of the single TSV interconnection, dual redundant TSV interconnection, and quad redundant TSV interconnection test structure design are 0.197, 0.538, and 0.998 dB, and reflection losses are all greater than 15 dB. Figs. 3.10 and 3.11, respectively, show the high-frequency performance of the dual-redundant TSV

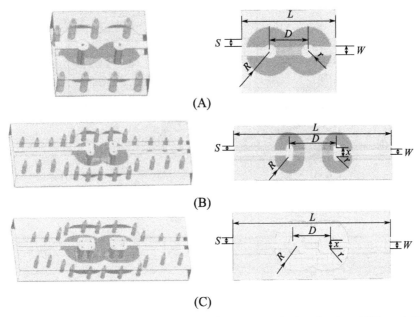

Fig. 3.8 HFSS model of test structures: (A) single TSV; (B) dual redundant TSV; and (C) quad redundant TSV.

Table 3.2 Dimensional parameters of redundant TSV test structure (unit: μm).

	L	S	W	D	R	r	x
Single TSV interconnection test structure	1000	70	100	400	250	75	–
Dual redundant TSV interconnection test structure	3000	70	100	400	250	75	160
Quad redundant TSV interconnection test structure	3000	70	100	640	250	75	120

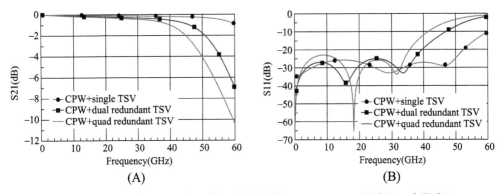

Fig. 3.9 S-parameter simulation results of the CPW+TSV test structure: (A) S21 and (B) S11.

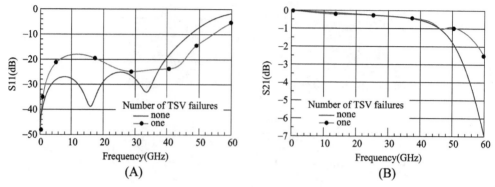

Fig. 3.10 Dual-redundant TSVs fault simulation results: (A) S11 and (B) S21.

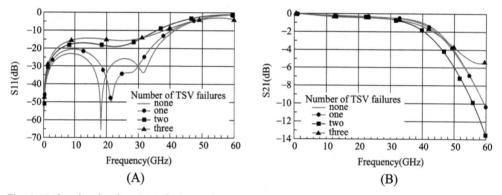

Fig. 3.11 Quad-redundant TSVs fault simulation results: (A) S11 and (B) S21.

and the quad-redundant TSV when different numbers of RF TSVs fail. It can be seen that as the number of failed TSVs increases, the resonance frequency increases, and the insertion loss decreases. When the frequency is lower than 44 GHz, the insertion loss change caused by TSV failure is small and can be ignored.

3.4 Sample processing and test result analysis

According to the developed process of the HR-Si interposer in our lab, the processed coaxial-like TSV and redundant TSV samples were fabricated, as shown in Fig. 3.12. Fig. 3.13 shows the high-frequency performance of RF TSVs with a grounding TSV ring with a changing diameter. For the best design, the insertion losses are less than 0.8 dB, and the larger the diameter of the ring-grounded TSV array, the smaller the insertion loss. The measured insertion loss of the single TSV, dual-redundant TSV, and quad-redundant TSV interconnection test structure at 40 GHz is 0.721, 1.18, and 1.635 dB,

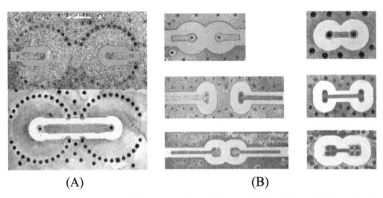

Fig. 3.12 Microscopic images of novel TSV samples: (A) coaxial-like TSV and (B) redundant TSV.

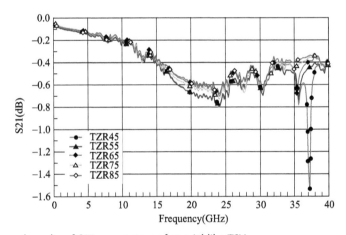

Fig. 3.13 Measured results of S21 parameters of coaxial-like TSV.

respectively. From the test results in Fig. 3.14, as the number of redundant TSVs increases, the insertion loss increases, and the trend is consistent with the simulation results.

However, the measurement results of the samples have a large deviation from the simulation results. In the frequency range of 0–40 GHz, the maximum deviations between simulated and measured insertion loss for single TSV, coaxial-like TSV, dual-redundant TSV, and quad-redundant TSV test structures are 0.53, 0.21, 0.84, and 0.95 dB, respectively. It is guessed that changes in the material parameters may occur during the manufacturing process, such as surface roughness and resistivity of copper, which are not factored into the simulation. Table 3.3 shows the resistivity of the copper layer measured by a four-probe tester, and its average resistivity is 12.79 μΩ·cm. Fig. 3.15 is the roughness of the copper layer measured by Profilometer, which shows that the roughness is about 60–70 nm, and some local areas reach about 150 nm due to oxidation.

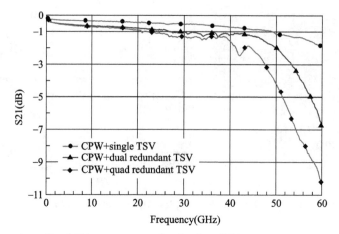

Fig. 3.14 Measured results of S21 parameters of redundant TSV.

Table 3.3 Resistivity measurement results.

Area	Resistivity ($\mu\Omega \cdot$ cm)
1	4.56
2	12.44
3	13.95
4	12.43
5	18.24
6	12.36
Average (Area 2–4, 6)	12.79

The roughness of the copper layer will cause a change of surface impedance, leading to an increase in conductor loss [14]. The influence of the surface roughness will become very significant, especially when the skin depth corresponding to the operating frequency is less than or equal to the surface roughness [15,16]. For decades, there has been a lot of research on the influence of roughness on RF performance, which can be divided into three main methods:

(a) Establish an equivalent model of the transmission structure and introduce a correction factor into the calculation of the circuit model.
(b) Obtain the topography of the metal surface, a three-dimensional structure with roughness details [17,18] is created and simulated using a commercial full-wave solver. This method has the problems of low computational efficiency and small modeling area.
(c) Calculate effective material parameters based on the electromagnetic field theory and substitute them into a three-dimensional field solver.

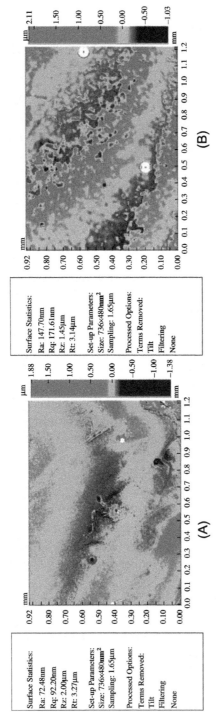

Fig. 3.15 Surface roughness test results: (A) areas with less roughness and (B) areas with large roughness.

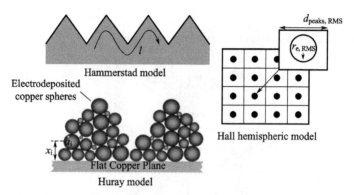

Fig. 3.16 Classical roughness model.

Regarding the first method, some empirical formulas based on hypothetical periodic structures have been proposed. The classical roughness model is shown in Fig. 3.16. The surface roughness of the Hammerstad [16] and Groisse models [19] is approximate to a zigzag pattern along the conductor surface. The limitations of the two models mean that they are only suitable for low-frequency or small root-mean-square (RMS) roughness conditions, because the maximum value of the correction factor when it reaches 2 will be saturated, which will lead to accuracy loss at high frequencies. The Hall hemispheric model [20] is an improvement over the Hammerstad model, but it usually overestimates the loss at low frequency and underestimates the loss at high frequency. Based on the hemisphere model, the Huray model [21] was introduced to model the surface roughness as a pyramid structure composed of stacked snowball particles on the surface of the conductor. The loss caused by surface roughness is equal to the sum of all snowball losses. However, the size of the snowball and the number of spheres in each protrusion are difficult to obtain. The representative of the third method is a surface roughness gradient model based on observable roughness parameters and electromagnetic field theory [22,23]. The model can predict the effect of roughness on loss and phase delay in a typical transmission line up to 100 GHz. However, it involves complex finite difference method (FDM) numerical calculations in the modeling process, which is not convenient in engineering applications.

In order to evaluate the influence of material parameters on RF performance, the monitored roughness and resistivity were substituted into the HFSS simulation. The roughness simulation uses the Groisse model because the coaxial-like TSV is essentially a single TSV interconnection; it is not simulated separately. The optimized simulation results are compared with the measured results in Fig. 3.17. It can be observed that the low-frequency deviation has been greatly reduced, while the maximum difference between the simulated and measured insertion loss for single TSV, dual-redundant

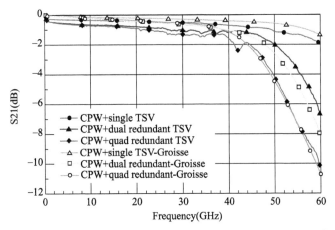

Fig. 3.17 Comparison of Groisse model simulation results and measurement results.

TSV, and quad-redundant TSV interconnection test structures are 0.5 dB at 50 GHz, 1.4 dB at 57 GHz, and 1 dB at 42 GHz, respectively. It is guessed that the correction factor of the Groisse model is saturated at high frequency.

Fig. 3.18 shows the detailed image of the surface profile of the copper layer obtained through a scanning electron microscope. The surface of the copper layer is formed by the accumulation of spherical particles, which conforms to the Huray snowball model. The correction factor of the Huray snowball model does not saturate at high frequency and has high accuracy. Based on the surface parameters of the copper layer obtained by SEM and surface profiler, as shown in Fig. 3.19, the snowball model is adopted to remodel and simulate. The measured peak height on the surface of copper layer is $h_{tooth} = 1.53\,\mu m$, the peak bottom width is $b_{base} = 34\,\mu m$, and the distance between the peaks is $d_{peaks} = 15\,\mu m$.

First, it is necessary to calculate the area of the protrusion. According to the SEM image, the shape of the protrusion can be approximated as a cone. Then the surface area can be calculated by the following formula:

$$A_{lat} = \frac{\pi b_{base}}{2}\sqrt{\left(\frac{b_{base}}{2}\right)^2 + (h_{tooth})^2} \qquad (3.1)$$

Substituting the measured value, the surface area is $9.11\,\mu m \times 102\,\mu m$. The radius of the sphere is 200 nm and the surface area of the individual sphere is $0.5\,\mu m^2$. The number of snowballs required to achieve the same surface area as the protrusion is:

$$N = \frac{A_{lat}}{A_{sphere}} = 1882 \qquad (3.2)$$

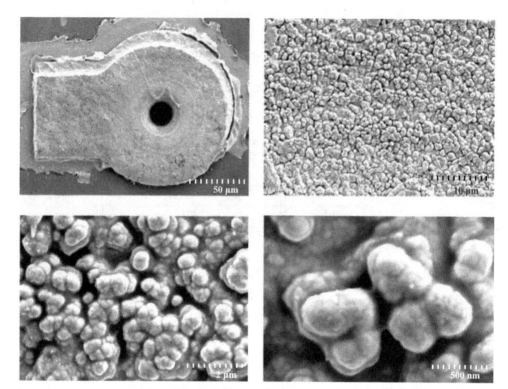

Fig. 3.18 SEM photo of surface copper layer.

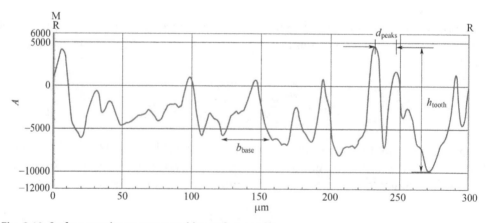

Fig. 3.19 Surface roughness measured by surface profiler.

According to the distance between the peaks, the area of the surrounding plane is calculated as:

$$A_{\text{title}} = d_{\text{peak2}}^2 = 225\,\mu m^2 \qquad (3.3)$$

To calculate the intrinsic impedance η, ε uses the dielectric constant value 11.9 of the silicon substrate under the copper:

$$\eta = \sqrt{\frac{\mu_0}{\varepsilon_0 \varepsilon_{Si}}} = 109\,\Omega \qquad (3.4)$$

$$k = \frac{2\pi}{\lambda} = \frac{2\pi\sqrt{\varepsilon_{Si}}}{c} \qquad (3.5)$$

The total power consumed by the roughness structure is the sum of the power consumed by N number of snowballs. The surface roughness correction factor is the ratio of the absorbed power with and without surface roughness, which can be obtained by the following formula:

$$K_{\text{Hurray}} = \frac{P_{\text{flat}} + P_{N \text{ spheres}}}{P_{\text{Plat}}} = \frac{(\mu_0 \omega \delta / 4) A_{\text{title}} + \sum_{n=1}^{N} \text{Re}\left[\eta(3\pi/k^2)(\alpha(1) + \beta(1))/2\right]_n}{(\mu_0 \omega \delta / 4) A_{\text{title}}} \qquad (3.6)$$

where the calculation formulas of $\alpha(1)$ and $\beta(1)$ can be found in reference [24]. The simulation results obtained by substituting the correction factor into the model are shown in Fig. 3.20. Compared with the Groisse model, the gap between the simulation results with Huray model and the measured results at high frequency is relatively narrow. However, as the frequency increases, the gap is still gradually increased. The maximum difference between the simulation and the measured insertion loss for the single TSV,

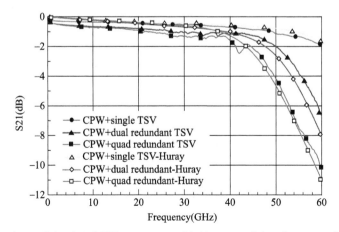

Fig. 3.20 Comparison of simulated S21 parameter with Huray model and measured S21 parameter.

dual-redundant TSV, and quad-redundant TSV interconnection test structure is 0.34 dB at 54 GHz, 1.3 dB at 57 GHz, and 0.95 dB at 42 GHz. At the same time, it can be observed that the simulation accuracy of the single TSV test structure is improved.

The simulation error is analyzed to be caused by the following: (a) The roughness of the whole structure is replaced by the roughness of the local area in the simulation process; (b) the modeling of metal protrusion is inaccuracy to a certain degree, which may be one of the reasons for the error; and (c) with the increase in frequency and the number of RF TSVs, the insertion loss of the whole structure is mainly contributed by TSV, but the surface roughness or random discontinuity of the Cu layer deposited on the side wall of TSV is difficult to distinguish and cannot be considered in the modeling at present. As a result, the high-frequency simulation results of the Groisse model and the Huray model of the redundant TSV are close.

To obtain the precise value of insertion loss contributed by RF redundant TSVs, de-embedding was conducted. According to the relevant theory of microwave network parameters, conversion into ABCD parameters with cascade characteristics for RF redundant TSV sample structures was carried out via parameter transformation and matrix operation [25]. The 1000-μm CPW test structure is viewed as four lengths of 250-μm CPW lines. A single RF TSV interconnect and redundant RF TSV interconnect test structure is viewed as three lengths of CPW lines and two TSV interconnect structures, as shown in Fig. 3.21. To simplify the description, J1, J2, J3, J4, and J5 are used to represent the CPW, single TSV interconnect, coaxial-like TSV interconnection, dual-redundant TSV interconnect, and quad-redundant TSV interconnect test structure. L1 represents 250-μm CPW and S-TSV represents a single TSV interconnect, L-TSV

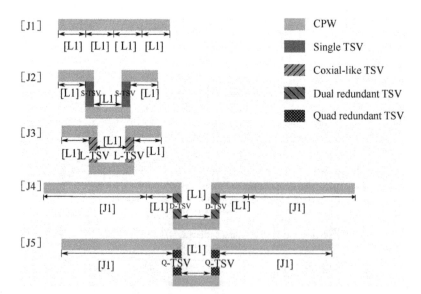

Fig. 3.21 Schematic diagram of test structure de-embedding.

represents a coaxial-like TSV interconnection, D-TSV represents a dual-redundant TSV interconnect, and Q-TSV represents a quad-redundant TSV interconnect structure.

The ABCD parameters corresponding to the five unit structures are represented by square brackets "[]" and the tag name. The ABCD parameters of the unit structure are multiplied by the unit structure to represent the ABCD parameters of the five test structures J1, J2, J3, J4, and J5 as

$$[J1] = [L1][L1][L1][L1] \quad (3.7)$$

$$[J2] = [L1][S-TSV][L1][S-TSV][L1] \quad (3.8)$$

$$[J3] = [L1][L-TSV][L1][L-TSV][L1] \quad (3.9)$$

$$[J4] = [J1][L1][D-TSV][L1][D-TSV][L1][J1] \quad (3.10)$$

$$[J5] = [J1][Q-TSV][L1][Q-TSV][J1] \quad (3.11)$$

The number of frequency points of the high-frequency measurement is marked as N, in the calculation process, set N-step loops to perform 2×2 matrix operations. The de-embedding process solves the loss contributed by a single TSV interconnect S-TSV, a coaxial-like TSV interconnection L-TSV, a dual redundant TSV interconnect D-TSV, and a quad-redundant TSV interconnect Q-TSV using the previous five matrix equations. Through the operations of square root and inversion matrix, the ABCD parameters matrix of each unit structure is obtained as

$$[L1] = [J1]^{\frac{1}{4}} \quad (3.12)$$

$$[S-TSV] = [L1]^{-1}\left([J2][L1]^{-1}\right)^{\frac{1}{2}} \quad (3.13)$$

$$[L-TSV] = [L1]^{-1}\left([J2][L1]^{-1}\right)^{\frac{1}{2}} \quad (3.14)$$

$$[D-TSV] = [L1]^{-1}\left([J1]^{-1}[J3][J1]^{-1}[L1]^{-1}\right)^{\frac{1}{2}} \quad (3.15)$$

$$[L2] = [L1][L1][L1] \quad (3.16)$$

$$[Q-TSV] = [L1]^{-1}\left([L2]^{-1}[J4][J1]^{-1}\right)^{\frac{1}{2}} \quad (3.17)$$

where [L2] is an intermediate variable for simplifying expressions. According to the transformation relationship between ABCD parameters and S-parameters, the ABCD parameters of each unit structure are converted into S-parameters, and the transmission characteristics of the three TSV interconnections can be compared. Microwave network parameter conversion and other operations in the de-embedding process are performed in MATLAB software.

Using the preceding TSV de-embedding method, the S-parameter of TSV interconnection can be obtained. S-parameters of coaxial-like TSV are shown in Fig. 3.22, and its

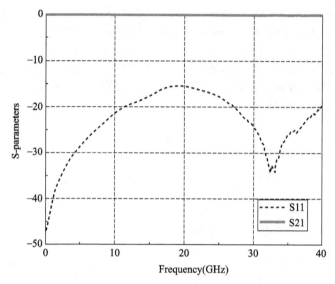

Fig. 3.22 S-parameters de-embedding results of coaxial-like TSV interconnection.

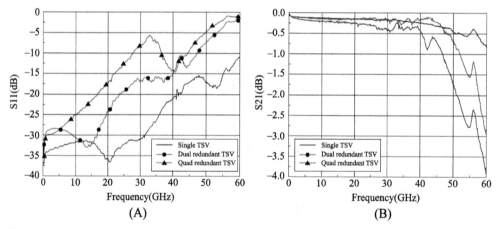

Fig. 3.23 S-parameters de-embedding results of redundant TSV interconnection: (A) S11 and (B) S21.

insertion loss is 0.07 dB at 40 GHz. The de-embedding results of redundant TSVs are shown in Fig. 3.23. The following points can be seen: (a) with the increase of frequency, the measured insertion loss value decreases gradually; (b) the measurement results and simulation results show a similar trend, the insertion loss increasing with the increase of the number of RF redundant TSVs; and (c) when the frequency is less than 40 GHz, the S21 values of the three TSV interconnects are close, and the difference is within 0.25 dB. When the frequency is greater than 40 GHz, the insertion loss of dual-redundant and quad-redundant TSV interconnects increases rapidly as the frequency

Table 3.4 Comparison RF TSV for high-frequency applications.

References	Substrate material	Type of vias	Transmission loss of one transition (dB) 10 GHz	Transmission loss of one transition (dB) 40 GHz	Via size (μm)	Via length (μm)
[26]	Glass	Single TGV	0.03	0.22	ϕ55	366
[27]	LCP	Single via	0.071	0.12	ϕ55	51
[28]	Si(HR)	Single TSV	0.05	–	ϕ100	300
[29]	Si(HR)	Single TSV	0.04	–	ϕ8 and ϕ90	25 and 280
[30]	Si(HR)	Single TSV	1.6	–	ϕ40	120
[31]	Si(HR)	Single TSV	0.37	–	ϕ10	100
This work	Si(HR)	Single TSV	0.11	0.22	ϕ40 and ϕ80	50 and 250
		Coaxial-like TSV	0.06	0.07		
		Dual-redundant TSV	0.14	0.19		
		Quad-redundant TSV	0.2	0.46		

increases, especially for quad-redundant TSV interconnect. These findings are sufficient to conclude that the redundant TSV design has the same RF capability as a single TSV. The RF performance of the TSV interconnection proposed in this work is compared with similar results published by other research teams in Table 3.4.

3.5 Optimization of HR-Si TSV interposer

According to the test results and analysis of RF TSV samples, the diameter and material parameters of TSV play a key role in RF performance. Among the factors, the resistivity and surface roughness of the electroplated copper layer should be more carefully treated. With organic additives in the electroplating solution for TSV metallization and residual glue adhering to the rough metal surface after photolithography, if the removal is not thorough, contamination occurs. An EDX test was carried out on the Cu surface of TSV samples. The element content showed a clear peak of carbon and oxygen, as shown in Fig. 3.24. Table 3.5 shows that the carbon content was 4.15% and the oxygen content was 2.32%. The presence of organic impurities can affect the conductivity of the metal, thereby affecting the radio frequency performance. To improve the RF properties, the effort toward optimization should not lie in the process step only, such as optimizing the

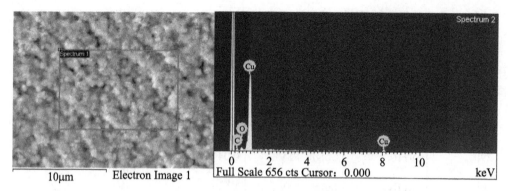

Fig. 3.24 EDX test results on the surface of the copper layer.

Table 3.5 Element content.

Element	Weight (%)	Atomic (%)
C (K^a)	4.15	17.60
O (K^a)	2.32	7.39
Cu (L^a)	93.53	75.00
Totals	100.00	~100.00

^{a}K, L, refer to the line system of characteristic X-rays.

electroplating process parameters or adopting copper CMP to form a smooth and continuing Cu layer with a reduced surface roughness, but also lies in the post-treatment step.

Here, surface treatment and an annealing experiment were tried to improve the RF property. A CPW line linked with RF TSVs was utilized as a test vehicle to do this research, and the layout is shown in Fig. 3.25. The measurement results of S21 parameters of samples with different CPW widths before and after surface treatment are shown in

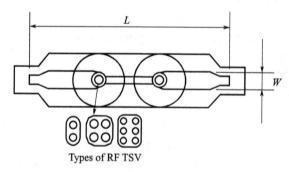

Fig. 3.25 Samples used for posttreatment.

Design, verification, and optimization of novel 3D RF TSV based on HR-Si interposer 85

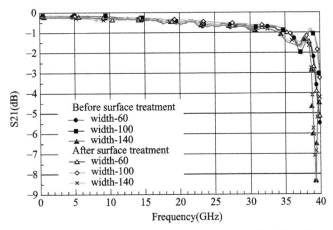

Fig. 3.26 S21 parameters of CPW with different line widths + single TSV before and after surface treatment.

Fig. 3.26. It can be seen that the high-frequency insertion loss of the sample decreases, and the large amount of insertion loss reduction is up to 0.5 dB around 30 GHz. As the line width of CPW increases, the reduction in insertion loss is greater, but it is not obvious. It is questioned whether it is caused by the short CPW length. Although the line width increases, the total area does not change much.

Surface treatment is performed with a group of CPW lines linked by redundant RF TSVs, and the measured S21 parameters are shown in Fig. 3.27. It can be seen that the variation of insertion loss of the test structure with the line length of 3 mm is three to six times that of the line length of 1 mm, which verifies that the larger the surface area of the

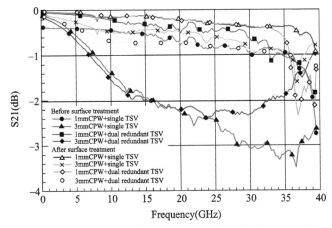

Fig. 3.27 S21 parameters of CPW with different line lengths + single TSV before and after surface treatment.

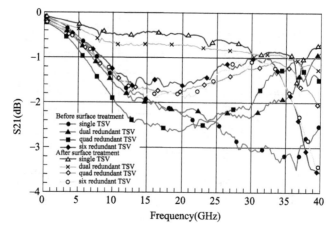

Fig. 3.28 S21 parameters of CPW and different numbers of RF TSVs before and after surface treatment.

test structure, the better the optimization effect of the surface treatment. At the same time, it is observed that in the frequency range of 30–40 GHz for dual-redundant TSV, the insertion loss decreases gradually slowly, and even reverses to increase. This phenomenon is not found on a single TSV structure. It is guessed that the optimization effect is related to the number of RF TSVs.

Fig. 3.28 shows the S-parameters before and after the treatment for CPW lines linked with different numbers of RF TSVs. As the number of RF TSVs increases, the insertion loss reduction after surface treatment gradually decreases. With regard to the number of RF TSVs, when it is 6, the insertion loss value does not change basically. Combining the characteristics of redundant TSV, as the number of RF TSVs increases, the greater the insertion loss; the high-frequency loss of the whole test structure is mainly contributed by the redundant TSV interconnection. The reason is that the surface treatment can only clean the planar structure but has less impact on the inside of TSVs, and the more redundant TSV numbers, the higher the uncertainty of the optimization effect in the hole. At the same time, the surface treatment will affect the surface condition of the copper layer to a certain extent, such as roughness change, oxidation, etc., causing the redundant TSV structure to increase the insertion loss when the frequency is about 40 GHz. The insertion loss changes of samples before and after surface treatment are totally summarized in Table 3.6 for reference.

A large amount of free energy is stored in the initial crystal grains of the copper layer produced by electroplating. Even at room temperature, the copper crystal grains will grow again, causing large internal tensile stress to form holes and increase the metal

Table 3.6 Change value of insertion loss before and after surface treatment RF TSV linked CPW lines.

Parameters		Value	Insertion loss change value (dB)				Maximum change value/frequency
			at 10 GHz	at 20 GHz	at 30 GHz	at 40 GHz	
CPW (length: 1 mm) + single TSV	Line width	60 μm	0.078	−0.12	−0.12	0.1	−0.18 dB at 30 GHz
		100 μm	0.07	−0.14	−0.16	0.04	−0.2 dB at 30 GHz
		140 μm	0.07	−0.13	−0.18	−0.4	−0.5 dB at 35 GHz
CPW (width: 100 μm) + TSV	Line length	CPW (length: 1 mm) + single TSV	−0.29	−0.54	−0.58	−2.4	−2.4 dB at 40 GHz
		CPW (length: 3 mm) + single TSV	−0.94	−1.8	−2.17	−1.82	−2.45 dB at 34 GHz
		CPW (length: 1 mm) + dual redundant TSV	−0.07	−0.26	−0.26	−1.4	−1.4 dB at 40 GHz
		CPW (length: 3 mm) + dual redundant TSV	−0.85	−1.5	−1.42	0.67	−1.6 dB at 24 GHz
CPW (length: 3 mm, width: 100 μm) + TSV	Number of RF TSVs	Single TSV	−0.9	−1.8	−2.1	−1.82	−2.45 dB at 34 GHz
		Dual-redundant TSV	−0.85	−1.5	−1.42	0.67	−1.6 dB at 24 GHz
		Quad-redundant TSV	−0.62	−0.9	−0.9	0.4	−1 dB at 24 GHz
		Six-redundant TSV	−0.02	0.2	0.03	−0.1	−0.13 dB at 28 GHz

Table 3.7 Metal surface roughness before and after annealing.

Temperature	Processed	Roughness (nm)				
		1	2	3	Average value	Change value
200 °C	Unannealed	48.6	68.7	66.2	61.2	–
	Annealed	52.4	58.7	61.7	57.6	−3.6
300 °C	Unannealed	83.8	69.7	67.4	73.6	–
	Annealed	62.7	66.8	81.2	70.2	−3.3
400 °C	Unannealed	45.3	83.7	62.8	63.9	–
	Annealed	73.2	62.3	45.1	60.2	−3.7
500 °C	Unannealed	61.1	66.7	67.1	64.9	–
	Annealed	56.3	60.6	62.7	59.9	−3.6
	Pickled	76.2	79.1	81.1	78.8	18.9

resistivity. At the same time, due to the mismatch of the coefficient of thermal expansion (CTE) between the copper layer in the TSV and the surrounding dielectric material, thermal reliability problems may be brought out [32–34]. Annealing is an efficient way to address the related problems of electroplating copper, as Cu film grains will regrow and release residual stress after annealing. With the growth of crystal grains, the grain boundary region becomes smaller and thinner, and the grain boundary potential barrier decreases, which is beneficial to the transport of carriers, thus increasing the mobility of carriers and reducing the resistivity of the copper layer [35].

Here, an annealing experiment was done with a CPW line linked by TSVs, which is convenient to extract material parameters. The annealing experiment was carried out in a N_2 environment to protect the copper from oxidation, and the annealing temperature was set from 200 to 500 °C. The heating rate was set to 15 °C per minute, the annealing duration was maintained for 30 min, and then the sample was cooled to room temperature under natural cooling conditions. AFM was used to measure the surface roughness of the copper layer before and after annealing. The results are shown in Table 3.7. The surface roughness of the annealed copper layer decreased slightly, but the sample annealed at 500 °C was severely oxidized. Subsequent electrical tests could not be carried out; after removing the oxide layer by pickling, the surface roughness was greatly increased. The resistivity of the copper layer was measured, and the comparison results before and after annealing are shown in Fig. 3.29. After annealing at 200, 300, 400, and 500 °C, the resistivity reduction ratio of the copper layer was 37.5%, 61%, 67%, and 73%, respectively; the resistivity reduction ratio increased with the increase of annealing temperature. The crystal grain morphology was investigated with SEM, and Fig. 3.30 shows the grains of the annealed copper

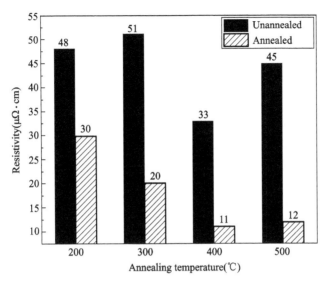

Fig. 3.29 Change in resistivity before and after annealing.

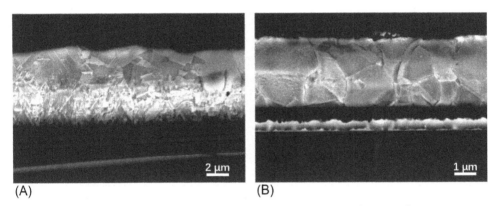

Fig. 3.30 SEM pictures of copper layer grains: (A) unannealed and (B) after annealing at 500°C.

layer regrowth and increase in size, thereby reducing the resistivity, which is in agreement with electrical test results.

Combining the previously described experiment, the RF transmission structure shown in Fig. 2.25 was annealed at temperatures ranging from 200 to 400°C. The measured S21 parameters before and after annealing are shown in Figs. 3.31 and 3.32, and the change values of the insertion loss are summarized in Table 3.8. The following conclusions can be drawn: (1) annealing treatment can improve the RF capability of the transmission structure, the main influence frequency band is 25–35 GHz, and the maximum

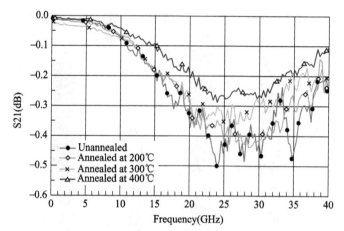

Fig. 3.31 S21 parameters of CPW before and after annealing.

reduction of insertion loss can reach 0.58 dB; (2) the higher the temperature in the range of annealing temperature from 200°C to 400°C, the smaller the insertion loss value; (3) under the same annealing temperature, as the number of RF TSVs increases, the overall insertion loss changes basically remain the same, which is different from the results of the surface treatment. It implies that the annealing treatment has an effect on the surface transmission structure and the vertical interconnection structure; and (4) the annealing temperature change has no obvious effect on the performance of the transmission structure containing TSV. The reason may be that the metal in the TSV changes shape due to the limitation of the surrounding silicon at high temperatures, which affects the signal transmission path, thereby increasing high-frequency loss.

3.6 Conclusion

This chapter discusses the design, fabrication, and characterization of a series of novel RF TSV interconnection structures based on a HR-Si interposer. The measurement results show that RF insertion loss for the proposed coaxial TSV, dual-redundant TSVs, and quad-redundant TSVs is 0.07 dB at 40 GHz, 0.14 dB at 40 GHz, 0.46 dB at 40 GHz, respectively. Combined with electrical testing and modeling analysis, it was found that the resistivity and surface roughness of metal are the main factors that affect the electrical performance. Optimization with surface treatment and annealing was done with CPW lines linked with RF TSVs, experimental validation was completed, and the optimization direction was concluded.

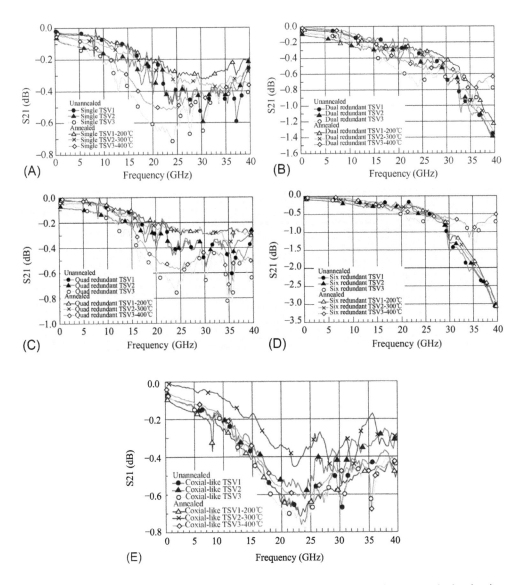

Fig.3.32 S21 parameters of CPW + TSV before and after annealing (A) single TSV, (B) dual redundant TSV, (C) quad redundant TSV, (D) six redundant TSV, and (E) coxial-like TSV.

Table 3.8 Change value of insertion loss before and after annealing.

Sample	Annealing temperature (°C)	Insertion loss change value (dB)				Maximum change value/frequency
		At 10 GHz	At 20 GHz	At 30 GHz	At 40 GHz	
CPW	200	−0.004	−0.07	−0.01	−0.01	−0.21 dB at 38 GHz
	300	0.006	−0.09	−0.12	−0.04	−0.29 dB at 38 GHz
	400	−0.029	−0.17	−0.16	−0.15	−0.3 dB at 38 GHz
CPW + single TSV	200	−0.012	−0.12	−0.22	−0.06	−0.29 dB at 37 GHz
	300	−0.06	−0.16	−0.24	0.04	−0.24 dB at 30 GHz
	400	−0.07	−0.14	−0.23	−0.05	−0.23 dB at 30 GHz
CPW + coaxial-like TSV	200	0.02	0.02	−0.1	0.03	−0.1 dB at 30 GHz
	300	−0.13	−0.2	−0.13	−0.005	−0.25 dB at 24 GHz
	400	−0.05	−0.12	−0.15	−0.004	−0.18 dB at 24 GHz
CPW + dual-redundant TSV	200	0.01	−0.17	−0.21	−0.14	−0.47 dB at 33 GHz
	300	−0.06	−0.16	−0.19	−0.01	−0.21 dB at 24 GHz
	400	−0.08	−0.17	−0.17	−0.14	−0.26 dB at 24 GHz
CPW + quad-redundant TSV	200	−0.02	−0.13	−0.21	−0.02	−0.35 dB at 36 GHz
	300	−0.07	−0.17	−0.2	0.02	−0.25 dB at 31 GHz
	400	−0.09	−0.12	−0.04	−0.14	−0.31 dB at 35 GHz
CPW + six-redundant TSV	200	0.01	−0.05	−0.3	−0.04	−0.58 dB at 34 GHz
	300	−0.07	−0.16	−0.22	−0.02	−0.22 dB at 30 GHz
	400	−0.1	−0.13	−0.18	−0.2	−0.48 dB at 36 GHz

References

[1] Ebefors T, Fredlund J, Perttu D, et al. The development and evaluation of RF TSV for 3D IPD applications. In: 2013 IEEE international 3D systems integration conference (3DIC). IEEE; 2013. p. 1–8.

[2] El Bouayadi O, Dussopt L, Lamy Y, et al. Silicon interposer: a versatile platform towards full-3D integration of wireless systems at millimeter-wave frequencies. In: 2015 IEEE 65th electronic components and technology conference (ECTC). IEEE; 2015. p. 973–80.

[3] Pares G, Jean-Philippe M, Edouard D, et al. Highly compact RF transceiver module using high resistive silicon interposer with embedded inductors and heterogeneous dies integration. In: 2019 IEEE 69th electronic components and technology conference (ECTC). IEEE; 2019. p. 1279–86.

[4] Xu Z, Lu JQ. Three-dimensional coaxial through-silicon-via (TSV) design. IEEE Electron Device Lett 2012;33(10):1441–3.

[5] Wang F, Wang G, Yu N. Electrical performance of coaxial ring TSV. J Comput Phys 2018;35(02):242–52 [in Chinese].

[6] Yook JM, Kim YG, Kim W, et al. Ultrawideband signal transition using quasi-coaxial through-silicon-via (TSV) for mm-wave IC packaging. IEEE Microwave Wireless Compon Lett 2020;30(2):167–9.

[7] Liu X, Chen Q, Dixit P, et al. Failure mechanisms and optimum design for electroplated copper through-silicon vias (TSV). In: 2009 59th electronic components and technology conference. IEEE; 2009. p. 624–9.

[8] Shen J, Chen P, Su L, et al. X-ray inspection of TSV defects with self-organizing map network and Otsu algorithm. Microelectron Reliab 2016;129–34.

[9] Kang U, Chung HJ, Heo S, et al. 8 Gb 3-D DDR3 DRAM using through-silicon-via technology. IEEE J Solid State Circuits 2009;45(1):111–9.

[10] Hsieh AC, Hwang TT. TSV redundancy: architecture and design issues in 3-D IC. IEEE Trans Very Large Scale Integr VLSI Syst 2011;20(4):711–22.

[11] Jiang L, Xu Q, Eklow B. On effective TSV repair for 3D-stacked ICs. In: 2012 design, automation & test in europe conference & exhibition (DATE). IEEE; 2012. p. 793–8.

[12] Vitale WA, Fernández-Bolaños M, Merkel R, et al. Fine pitch 3D-TSV based high frequency components for RF MEMS applications. In: 2015 IEEE 65th electronic components and technology conference (ECTC). IEEE; 2015. p. 585–90.

[13] Rahimi A, Somarajan P, Yu Q. Modeling and characterization of through-silicon-Vias (TSVs) in radio frequency regime in an active interposer technology. In: 2020 IEEE 70th electronic components and technology conference (ECTC). IEEE; 2020. p. 1383–9.

[14] Chen CD, Tzuang CKC, Peng ST. Full-wave analysis of a lossy rectangular waveguide containing rough inner surfaces. IEEE Microwave Guided Wave Letters 1992;2(5):180–1.

[15] Palasantzas G. Influence of self-affine and mound roughness on the surface impedance and skin depth of conductive materials. J Phys Chem Solid 2004;65(7):1271–5.

[16] Scogna AC, Schauer M. Performance analysis of stripline surface roughness models. In: 2008 International symposium on electromagnetic compatibility-EMC Europe. IEEE; 2008. p. 1–6.

[17] Li Q, Shi J, Chen KS. A generalized power law spectrum and its applications to the backscattering of soil surfaces based on the integral equation model. IEEE Trans Geosci Remote Sens 2002;40(2):271–80.

[18] Sain A, Melde KL. Surface roughness modeling for CB-CPWs. In: Proceedings of the 2012 IEEE international symposium on antennas and propagation. IEEE; 2012. p. 1–2.

[19] Groiss S, Bardi I, Biro O, et al. Parameters of lossy cavity resonators calculated by the finite element method. IEEE Trans Magn 1996;32(3):894–7.

[20] Hall S, Pytel SG, Huray PG, et al. Multigigahertz causal transmission line modeling methodology using a 3-D hemispherical surface roughness approach. IEEE Trans Microwave Theory Tech 2007;55(12):2614–24.

[21] Huray PG, Hall S, Pytel S, et al. Fundamentals of a 3-D "snowball" model for surface roughness power losses. In: 2007 IEEE workshop on signal propagation on interconnects. IEEE; 2007. p. 121–4.

[22] Gold G, Helmreich K. A physical surface roughness model and its applications. IEEE Trans Microwave Theory Tech 2017;65(10):3720–32.

[23] Chen L, Tang M, Mao J. An analytical gradient model for the characterization of conductor surface roughness effects. In: 2018 IEEE/MTT-S international microwave symposium-IMS. IEEE; 2018. p. 1036–8.

[24] Hall SH, Heck HL. Advanced signal integrity for high-speed digital designs. Wiley; 2009. p. 249–95.
[25] Kinayman N, Aksun MI. Modern microwave circuits. Artech House; 2008. p. 58–68.
[26] Khan WT, Tong J, Sitaraman S, et al. Characterization of electrical properties of glass and transmission lines on thin glass up to 50 GHz. In: 2015 IEEE 65th electronic components and technology conference (ECTC). IEEE; 2015. p. 2138–43.
[27] Chung DJ, Bhattacharya SK, Papapolymerou J. Low loss multilayer transitions using via technology on LCP from DC to 40 GHz. In: 2009 59th electronic components and technology conference. IEEE; 2009. p. 2025–9.
[28] Chen B, Sekhar VN, Jin C, et al. Low-loss broadband package platform with surface passivation and TSV for wafer-level packaging of RF-MEMS devices. IEEE Trans Compon Packag Manuf Technol 2013;3(9):1443–52.
[29] Ebefors T, Fredlund J, Perttu D, et al. The development and evaluation of RF TSV for 3D IDP applications. In: IEEE International 3D Systems Integration Conference(3DIC). IEEE, 3013; 2013. p. 1–8.
[30] Lorival JE, Calmon F, Sun F, et al. An efficient and simple compact modeling approach for 3-D interconnects with IC's stack global electrical context consideration. Microelectron J 2015;46(2):153–65.
[31] Kim N, Wu D, Kim D, et al. Interposer design optimization for high frequency signal transmission in passive and active interposer using through silicon via (TSV). In: 2011 IEEE 61st electronic components and technology conference (ECTC). IEEE; 2011. p. 1160–7.
[32] Kim BJ, Kim JH, Hwang SH, et al. Microstructure evolution and defect formation in cu through-silicon vias (TSVs) during thermal annealing. J Electron Mater 2012;41(4):712–9.
[33] Jiang QT, Frank A, Havemann RH, et al. Optimization of annealing conditions for dual damascene Cu microstructures and via chain yields. In: 2001 Symposium on VLSI technology. Digest of technical papers (IEEE cat. No. 01 CH37184). IEEE; 2001. p. 139–40.
[34] Okoro C, Vanstreels K, Labie R, et al. Influence of annealing conditions on the mechanical and microstructural behavior of electroplated Cu-TSV. J Micromech Microeng 2010;20(4), 045032.
[35] Luo X, Zhao H, Wu X. Effect of annealing temperature on microstructure and resistivity of Cu films. Semiconductor Tech 2008;(01). 77–79+89 [in Chinese].

CHAPTER 4

HR-Si TSV integrated inductor

4.1 Introduction

TSV interposer integrated passive devices (IPDs) are the natural technical options for TSV 3D RF integration module passive components. This term refers to the integrated production of substrate and passive components achieved by performing a MEMS process and semiconductor process on a silicon or glass interposer. The minimum line width (<10 μm magnitude) and high component parameter control accuracy of an IPD favor improvements in passive component performance and system integration level. An inductive element is an indispensable basic component in a RF system. As the miniaturization and integration of RF systems are continuously developing, the HR silicon TSV integrated inductor has become an important development direction.

Research on silicon-based integrated inductors has a long history, and rich research results, involving theoretical models, new materials, structure design and process, testing and other topics, have been reported. As effects due to the silicon substrate are complex, a high-precision theoretical model is required as an indispensable support tool that can save development time and cost for the whole system. In early related studies, the calculation method of inductor quality factor and inductance, for example, the lumped parameter model of the classical integrated inductor proposed by Patir CkuYe of Stanford University, is usually not comprehensive enough in terms of loss factors (e.g., eddy current, proximity effect) and therefore yields results deviating from experimental results. On the basis of Patir CkuYe's model, Hubert Curien (University of Lyon's Laboratory) and Ren-Jia Chan et al. presented models that take the proximity effects and substrate losses into consideration [1–5]. However, most of these studies still do not consider the roughness of the metal surface and substrate surface caused by the actual process, which does have a nonnegligible impact on inductor parameters.

Many new materials, new processes, and new structures have also been applied in the manufacture of inductors. In 2012, Wolfgang A. Vitale, a researcher at EPFL Nanoelectronics Laboratory, achieved the establishment of a flat spiral inductor on the surface of an HR-Si substrate [6]. In 2018, the U.S. Army Research Laboratory proposed a method to make low-cost 3D components by adopting a bending technique. Based on the local thermal effect of lasers, the material will deform under a thermal stress, such that a 3D inductor can be fabricated without any manual operation. A planar inductor produced in this manner has an inductance of 199 nH and a peak quality factor of 99.7 at 54.7 MHz. For a ring coil produced in this manner, the inductance is 159 nH and the peak quality factor is 64.9 at 86.7 MHz. On the basis of a traditional planar inductor, researchers from Fudan University used silicon oxide as a medium layer and achieved a high Q value and high inductance on the surface of a traditional Si substrate by adopting a vertical stacking structure. Simulation results show that at 7 GHz, the Q value of the structure can reach 32, and its inductance can reach 4 nH. Compared to the traditional planar inductor, the Q value of the new structure was increased by 30%. In Federal Polytechnic of Lausanne in Switzerland, taking advantage of vanadium dioxide's nature of converting from an insulator to metal, researchers placed switches made of molybdenum dioxide (MoO_2) in the winding space, and established a reconfigurable planar inductor on an HR-Si substrate. An inductor manufactured using this method can operate in the range of 4–10 GHz. For an operating frequency of 5 GHz, its inductance is 2.1 nH at 20°C, and 1.35 nH at 100°C. The inductor has a reconfiguration ratio of up to 55%, while it also has good stability throughout the frequency band (4–10 GHz). For an operating frequency of 7 GHz, its quality factor is about 8 at 20°C (off state), and 3 at 100°C (on state). However, due to the semiconductor characteristics of Si, it faces the problem of high-frequency loss of substrate, which can result in low Q. By adopting new structures [7–10], the quality factors of some inductors have reached 30 at C band. However, new structures, such as reverse groove, are usually complex and can introduce further technical risks and errors. TSV technology offers new opportunities for the development of silicon-based integrated inductors. Paragkumar A. Thadesa and other researchers at Georgia Tech have achieved a spiral inductor consisting of copper TSV interconnections and rewiring layers inside the cavity of a low-resistivity silicon (resistivity = 10 Ω/cm) substrate, by performing a SU sacrificial layer process. The Cu coil has a thickness of 8–10 μm, line width of 55–65 μm, inductance of 1.14 nH, and Q factor of 55 at 6.75 GHz. Researchers at Silex Microsystems, CEA-Leti, and other institutions manufactured an on-substrate 3D spiral inductor substrate made up of copper TSV interconnects and rewiring layers on an HR-Si TSV interposer, which has inductance of 1.4–12 nH and Q value of 10–100 [11]. Compared to the traditional new structure silicon-based integrated inductor, the HR-Si TSV interposer integrated inductor is directly composed of a Cu TSV interconnection and rewiring layer. It has high process compatibility and high integration capability as well as excellent device performance. This chapter focuses on the research progress of our research group in HR-Si TSV interposer integrated inductors, including a theoretical model and device design of a planar inductor as well as a TSV 3D inductor (Table 4.1).

4.2 HR-Si TSV interposer integrated planar inductor

In traditional on-chip integrated planar spiral inductors, semiconductor processes are usually used to establish a metal structure on the surface of the substrate as well as an

Table 4.1 Recent research results of silicon-based integrated inductors.

References	U.S. Army [12]	Fudan University [13]	EPFL [14]	CUNY [15]	KETI [16]	CAS [17]
Date	2018.5	2018.5	2018.8	2012.11	2012.5	2015.10
Operating frequency (GHz)	<1	>10	4–10	<20	—	<10
Q value	99.7 at 54.7 MHz	32 at 7 GHz	10.8 at 7 GHz at 20°C	9.1 at 2 GHz	40 at 2 GHz	33 at 5.5 GHz
Inductance (nH)	199	4	2.1–1.35 at 5 GHz	4.52 at 2 GHz	2.5 at 2 GHz	2.5 at 5.5 GHz
Substrate material	Air	Si	Si	Si	Si	—
Dimensions	—	—	—	$<1.5 \times 1.5 \times 0.8\,mm^3$	—	—

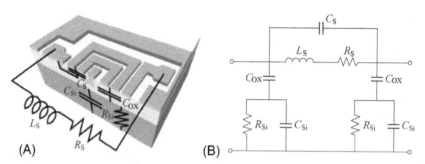

Fig. 4.1 Two theoretical models of planar spiral inductors: (A) Physical model of planar spiral inductor; (B) Lumped parameter model of planar spiral inductance [18].

underbridge structure that works as the second port of a two-port element, as shown in Fig. 4.1A. The lumped parameter model is shown in Fig. 4.1B, where L_S is the self-inductance of the inductor coil when it is assumed to be an ideal inductor, R_S is the self-impedance of the metal spiral inductor introduced by the deviation of the real inductor from the ideal inductor, C_S is the feed-through capacitance between the metal spiral coil and the lower lead metal, C_{OX} is the insulation layer capacitance between the metal spiral coil and the bottom of the insulation layer, C_{Si} is the coupling capacitance of Si substrate, and R_{Si} is the leakage resistance corresponding to C_{Si} [19].

The quality factor Q is an important indicator for the design of inductors, which to some extent reflects their ability to store electromagnetic (EM) field energy. Reducing the loss of EM field energy can effectively improve the quality of the inductive structure. The substrate and conductor materials are the main factors that affect the quality factor Q of the inductor. The losses associated with substrate material are generally due to leakage and eddy current, while the losses associated with the conductor material are mainly ohmic loss and eddy current loss. Regarding the substrate loss of classic planar spiral inductors, subsequent analysis will be from the perspective of structure and substrate material. Fig. 4.2 shows an integrated planar spiral inductor on an HR-Si TSV interposer, in which TSV interconnections are used to establish the underbridge structure. In order

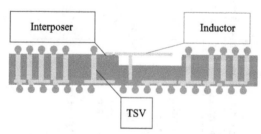

Fig. 4.2 Schematic diagram of integrated floating plane spiral inductor on HR-Si TSV interposer.

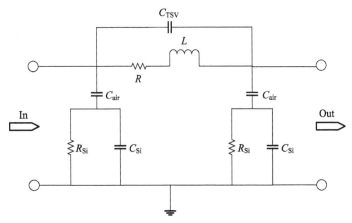

Fig. 4.3 Equivalent circuit model of integrated planar inductor on HR-Si TSV interposer.

to explore the mechanism of the substrate introducing loss, a cavity substrate has been formed by performing local RIE on the substrate, such that the planar spiral inductor is suspended to reduce substrate loss and improve the quality factor Q of the inductor. The lumped parameter circuit model is shown in Fig. 4.3.

As shown, the C_{TSV} is the feed-through capacitance between TSV and the planar spiral structure, L is the inductance, R is the comprehensive resistance of the conductor under the influence of various effects, C_{air} is the coupling capacitance introduced by the cavity structure, C_{Si} is the coupling capacitor of remaining silicon substrate under the cavity structure, and R_{Si} is the leakage resistance corresponding to C_{Si}. The parasitic parameters are calculated as follows:

L: The self-inductance of the planar spiral inductor is calculated using the Greenhouse method [20], which sums up the self-inductance of every straight wire and the mutual inductance with other parallel wires. Eq. (4.1) is total self-inductance of an inductor:

$$L_s = \sum_{i=1}^{N} L_i + \sum_{i=1}^{N} \sum_{\substack{i=1 \\ j \neq i}}^{N} M(i,j) \qquad (4.1)$$

where N is the total count of segments of the planar spiral inductor; i, j are the numbers of each segment; L_i corresponds to the self-inductance of each segment; $M(i,j)$ is the mutual inductance between the ith segment of straight wire and the jth segment of straight wire, either positive or negative.

L_i can be calculated by Eq. (4.2):

$$L_i = 0.002 l_i \left[\ln\left(\frac{2l_i}{GMD}\right) - 1.25 + \frac{AMD}{l_i} + \left(\frac{\mu}{4}\right) T \right] \qquad (4.2)$$

Table 4.2 The relationship between frequency correction factor T and frequency.

T	Film thickness	Frequency (GHz)
0.9974	10,000 Å (1 Å = 10^{-10} m)	10
0.9986	0.0025 µm (~0.1 mil, 1 mil = 25.4 µm)	1
0.9095	0.0075 µm (~0.3 mil)	1

where l_i is the length of the ith segment of straight wire; μ is the magnetic permeability of the planar spiral inductor coil; GMD is the geometric mean distance of the rectangular cross section of the inductor [21], and AMD is the algebraic average distance. For rectangular sections, GMD and AMD can be calculated by Eq. (4.3):

$$\text{GMD} = 0.2232(t+w), \quad \text{AMD} = \frac{t+w}{3} \quad (4.3)$$

where t is the thickness of the metal spiral coil; w is the line width of the metal spiral coil, as shown in Fig. 4.4. From Table 4.2, the relationship between frequency correction factor, line thickness, and frequency can be acquired, while the value of frequency correction factor in RF range is approximately 1.

When calculating $M(i,j)$, if two segments of wire are geometrically orthogonal to each other, the mutual inductance between the two is simply 0 (and therefore negligible). However, it becomes nontrivial if two segments are in parallel. As shown in Fig. 4.5, the ith and jth segments are straight wire segments of different lengths, where L_i and L_j are the lengths of the segments; p and q are distances between the ends of j and the ends of i; and GMD_M is the geometric mean distance between i and j, which is equal to the distance between the center lines of the two segments. GMD_M can be calculated by Eq. (4.4).

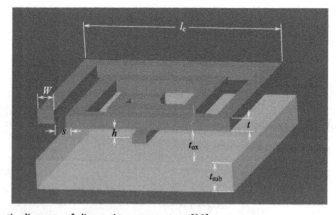

Fig. 4.4 Schematic diagram of dimension parameters [22]

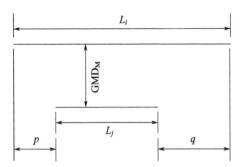

Fig. 4.5 Schematic diagram of GMD.

$$l_n(\text{GMD}_M) = \ln(d) - \left[\frac{1}{12\left(\frac{d}{w}\right)^2} + \frac{1}{60\left(\frac{d}{w}\right)^4} + \frac{1}{168\left(\frac{d}{w}\right)^6} + \ldots \right] \quad (4.4)$$

where d is the distance between center lines of two parallel straight metal wires m and j, and w is the width of the metal wires.

Different situations of current direction can yield different mutual inductance $M(i,j)$. When the current in two metal wires flows in the same direction, the mutual induction is positive and the inductance can be calculated by Eq. (4.5):

$$M(i,j) + M(j,i) = \left(M_{L_j+p} + M_{L_j+p}\right) - \left(M_p + M_q\right) \quad (4.5)$$

When the current in two metal wires flows in the opposite direction, the mutual induction is negative and the inductance should be calculated by Eq. (4.6):

$$M(i,j) + M(j,i) = \left(M_p + M_q\right) - \left(M_{L_j+p} + M_{L_j+q}\right) \quad (4.6)$$

The M_X terms in Eqs. (4.5) and (4.6) can be calculated by Eq. (4.7):

$$M_X = 2XU_X \quad (4.7)$$

where U_X is the mutual inductance parameter, as described in Eq. (4.8):

$$U_X = \ln\left[\frac{X}{\text{GMD}_M} + \sqrt{1 + \left(\frac{X}{\text{GMD}_M}\right)^2}\right] - \sqrt{1 + \left(\frac{\text{GMD}_M}{X}\right)^2} + \frac{\text{GMD}_M}{X} \quad (4.8)$$

So far, inductance due to the planar spiral inductor itself, which is only related to the size and the material of the coil, is determined. The real inductor is integrated on an Si substrate, and there are other parasitic parameters that can cause the total inductance of the metal spiral inductor to show different values at different frequencies.

Subsequently, the calculation method for the parasitic parameters in the theoretical model will be presented.

R: The impedance of the conductor itself is the sum of impedances due to many factors, mainly including DC impedance, skin effect, proximity effect, and surface roughness.

The DC impedance of the conductor itself is as Eq. (4.9) [23]:

$$R_{DC} = \frac{\rho l}{S} \tag{4.9}$$

As the electrical signal frequency increases, the skin effect of the current inside the metal becomes more and more significant, in the form of an increasing R_S [24–26]:

$$R_S = \frac{\rho l}{w\delta\left(1 - e^{-\frac{t}{\delta}}\right)} \tag{4.10}$$

where ρ is the resistivity of the metal; l is the total length of the center lines of all segments composing the metal spiral coil; w is width of the metal wire; t is thickness of the metal spiral coil; δ is the skin depth at high frequency. Skin depth δ can be calculated by Eq. (4.11):

$$\delta = \sqrt{\frac{2}{\omega\mu\sigma}} \tag{4.11}$$

where μ is the permeability values, σ is the electrical conductivity of the metal material, and ω is the angular frequency of the electrical signal.

Compared to DC conduction, in a metal conductor, the skin effect can dramatically reduce the cross-sectional area of effective electrical conduction, resulting in a sharp increase in the impedance loss of the conductor itself. The impact of skin effect on impedance is not significant in the low-frequency range, where the skin depth is greater than the thickness of the metal conductor and the area of effective conduction is not reduced substantially. However, at RF or higher frequencies, the impedance of the conductor itself will rocket due to the increase in impedance loss caused by the skin effect.

In a real inductor, due to the small spacing between neighboring metal segments, the magnetic field excited by the high-frequency signal in one metal segment will yield an eddy current in another, which can further reduce the effective conduction area by changing the current distribution [25,27]. Such a mechanism of in-metal eddy current loss is shown in Fig. 4.6.

This article considers that in each metal segment, loss P_n of the metal consists of two parts, which are the impedance loss $I_{ex}^2 R_n$ associated with the excitation current I_{ex}, and the impedance loss $I_{eddy}^2 R_{eddy_n}$ associated with the eddy current I_{eddy}. In this scenario, the loss power will be:

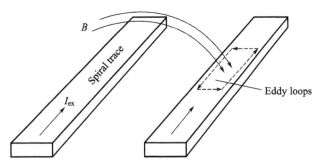

Fig. 4.6 Schematic diagram of the mechanism of eddy current loss.

$$P_n = I_{ex}^2 R_n + I_{eddy}^2 R_{eddy_n} \qquad (4.12)$$

where R_n and R_{eddy_n} are the DC impedance and eddy current impedance of the nth metal segment, respectively. R_n can be split and rewritten in the form of:

$$R_n = R_{sheet} \frac{l_n}{W} \qquad (4.13)$$

where l_n is the length, W is the width, R_{sheet} is the unit resistance after splitting. So it can be said that:

$$R_{eddy_n} \approx 2 R_{sheet} \frac{l_n}{\frac{W}{4}} \qquad (4.14)$$

Thus DC impedance can be used for the equivalence of eddy current loss. In real fabrication, after surface etching, the segment of metal spiral on the surface may bend due to the stress inside. For such a situation, the current in one segment will be reduced by surface excitation due to the magnetic field components in neighboring metal segments. Furthermore, the proximity effect will also be weakened. Therefore, on the basis of previous results, a coefficient γ should be introduced to characterize the resistance γR_{sheet} associated with the proximity effect after surface bending.

Considering the roughness of metal and nonmetal surfaces of real samples, when the skin depth (of the skin effect) is less than or comparable to the surface roughness in the skin effect, the effect due to surface roughness should be considered. Fig. 4.7 shows the test result of the surface roughness of the copper layer, using an optical surface analyzer. The protrusions have an average size of more than 0.5 µm and therefore a large area equivalent surface. This means the equivalent surface cannot be neglected because the path of current has been lengthened. If protrusions exist, current will flow along the rough surface in a longer path (Fig. 4.8), and thus the loss will be increased and the general quality factor will be decreased. Therefore the following section discusses how to account for the surface roughness in the theoretical model.

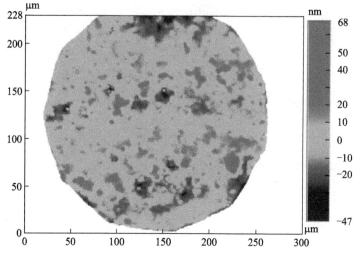

Fig. 4.7 Optical test results of surface roughness.

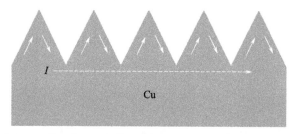

Fig. 4.8 Schematic diagram of surface roughness increasing current conduction distance.

Regarding the surface roughness, a modified Huray model can be used for approximation. The recent Huray model is effective for modeling rough surfaces [28,29]. The model is based on observations of real rough surfaces and uses a spherical structure to simulate a rough structure. It ensures that the rough surface area is the same, so the equivalence of real surface roughness can be achieved. The surface roughness factor calculated by using the Huray model is:

$$K_{\text{Huray}} = \frac{P_{\text{flat}} + P_{\text{spheres}}}{P_{\text{flat}}} \quad (4.15)$$

To improve the model, a factor τ that varies with frequency is introduced, since in real situations the path of current flowing through the conductor surface is not the same at different frequencies. In the low-frequency range, the small roughness of the metal surface will not affect the current path. In the case of high frequency, the current will be under the influence of skin effect together with other effects. The modified Huray model can better reproduce the test results:

$$K_{\text{Huray}} = \frac{P_{\text{flat}} + P_{\text{spheres}}}{P_{\text{flat}}} \tau \quad (4.16)$$

where τ is

$$\tau = \frac{\omega - \omega_{Q_{\max}}}{\omega_{\max}} + \frac{K_{\text{Huray}_{\max}}}{4} \quad (4.17)$$

where $\omega_{Q_{\max}}$ is the frequency when quality factor is maximum; $K_{\text{Huray}_{\max}}$ is the maximum surface roughness factor obtained in the simulation frequency range using a classic Huray model.

So far, three factors of losses that impact traditional planar spiral inductors are given: skin effect, proximity effect, and surface roughness. By integrating the factored into the original series resistance of a metal inductor, a modified expression of series resistance can be obtained:

$$R = R_{\text{DC}} + R_{\text{skin}} + R_{\text{eddy}} + R_{\text{roughness}} \quad (4.18)$$

The four terms on the right side of the equation are the DC impedance, the impedance due to skin effect, the impedance due to the proximity effect, and the impedance due to surface roughness, respectively.

C_S: C_S is the capacitance of the parasitic capacitor between the metal planar spiral inductor and TSV, which is connected in parallel with the inductor and constructs a lossy LC parallel resonant circuit. This circuit can directly impact the resonant frequency. If the thickness of underlying metal is neglected, the approximation of C_S can be figured out by referring to the model of the parallel plate capacitor. Define S as the equivalent area of the capacitor formed by the horizontal plane spiral inductor and the vertical TSV structure. The equation will be:

$$C_S = S \frac{\varepsilon_{\text{ox}}}{t_{\text{ox}}} \quad (4.19)$$

where ε_{ox} is the dielectric constant of the insulator material, and t_{ox} is the thickness of the insulator material between the lower surface of the metal spiral and the upper surface of the lower layer lead out metal.

C_{OX}: Because the silicon substrate is not an ideal insulator, under the induction of a metal inductor, an electrical charge will be accumulated on the upper surface of the silicon substrate. While the insulation layer can work as a medium, it will form a capacitor between the coil and the silicon substrate. Define l as the equivalent length of the wire and w as the equivalent width of the wire. The capacitance of such a capacitor can be calculated by:

$$C_{\text{OX}} = \frac{1}{2} lw \frac{\varepsilon_{\text{ox}}}{t_{\text{ox}}} \quad (4.20)$$

C_{Si}: Similarly, due to the semiconductor nature of silicon, on the lower surface of the silicon substrate corresponding induced charges will be accumulated corresponding to the induced charge on the upper surface. A substrate coupling capacitor will be formed, for which the capacitance is:

$$C_{Si} = \frac{1}{2} lw C_{Sub} \qquad (4.21)$$

R_{Si}: Following the previous discussion, due to the semiconductor nature of silicon, leakage loss will be introduced in the formation of the silicon substrate coupling capacitor. Such loss is critical for the inductor Q value, which can be calculated by:

$$R_{Si} = \frac{2}{lw C_{Sub}} \qquad (4.22)$$

Simplify the single-π model in Fig. 4.3 and an equivalent network model circuit can be obtained, as shown in Fig. 4.9, where Y_S and Y_{Sub} are calculated by Eqs. (4.23) and (4.24), respectively.

$$Y_S = \frac{1}{R_S + j\omega L_S} + j\omega C_S \qquad (4.23)$$

$$Y_{Sub} = \left(\frac{1}{R_{Si}} + jC_{Si}\right)/(jC_{OX}) \qquad (4.24)$$

Setting the left side as port 1 for the network and the right side as port 2, a parameter matrix can be obtained:

$$\begin{pmatrix} Y_{11} & Y_{12} \\ Y_{21} & Y_{22} \end{pmatrix} = \begin{pmatrix} Y_S + Y_{Sub} & -Y_S \\ -Y_S & Y_S + Y_{Sub} \end{pmatrix} \qquad (4.25)$$

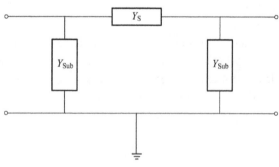

Fig. 4.9 Equivalent network model.

Finally, the corresponding expressions of quality factor Q and inductance L are calculated as follows:

$$Q = \frac{\text{Im}\left(\dfrac{1}{Y_{11}}\right)}{\text{Re}\left(\dfrac{1}{Y_{11}}\right)} \qquad (4.26)$$

$$L = \frac{\text{Im}\left(\dfrac{1}{Y_{11}}\right)}{\varpi} \qquad (4.27)$$

So far, all the calculations associated with the theoretical model of planar spiral inductor, as well as the calculation associated with two important parameters of inductors, have been given. As long as the design of any planar spiral inductor is known, the corresponding mathematical analysis can be carried out by using Matlab software and HFSS software on the basis of the model and the expressions of each parameter indicated.

Planar spiral inductors can be designed as different shapes according to specific requirements. While circular winding and rectangular winding are the most common cases, other windings, such as regular octagons, do exist (Fig. 4.10). Based on the discussed theoretical model as well as the lumped parameter mode, the Q value and inductance value can be calculated. Furthermore, the impact due to process-related parameters on the quality factor Q value and inductance can also be analyzed.

In order to verify the validity of the new structure proposed in this chapter, inductor structures of two sizes are designed as test samples. For both of them, the common dimension parameters are: number of turns is 2, line width is 40 μm, and turn spacing is 10 μm.

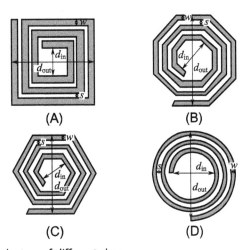

Fig. 4.10 Planar spiral inductors of different shapes.

Table 4.3 Matlab simulation parameter value.

Parameter	Value	Parameter	Value
Simulation frequency range (GHz)	1–20	Coil permeability (H/m)	$4\mu_0$
Total length of inductor (µm)	1803	Coil conductivity (S/m)	2.8×10^7
Inner diameter length (µm)	135	Coil dielectric constant (F/m)	ε_0
Coil number of turns	2.5	Silicon substrate permeability (H/m)	μ_0
Coil width (µm)	15	Silicon substrate conductivity (s/m)	10
Coil thickness (µm)	2	Silicon substrate dielectric constant (F/m)	ε_0
Turn spacing (µm)	1.5	Oxide magnetic permeability (H/m)	μ_0
Oxide layer thickness (µm)	4.8	Oxide layer conductivity (s/m)	0
Dielectric substrate thickness (µm)	300	Dielectric constant of oxide (F/m)	$4\varepsilon_0$

Where vacuum permeability $\mu_0 = 4\pi \times 10^{-7}$ H/m, and vacuum dielectric constant $\varepsilon_0 = 8.854187817 \times 10^{-12}$ F/m

The inductor A (named N2W40S10L400) has an outer diameter of 400 µm, while the inductor B (named N2W40S10L600) has an outer diameter of 600 µm [30,31]. The parameters used in the calculation are summarized in Table 4.3. Fig. 4.11 shows the result of computation for the Q value and inductance. Figs. 4.12 and 4.13 show the influence of dimension parameters (e.g., number of turns, coil width, thickness, turn spacing) on Q value and inductance, respectively. The qualitative trends are summarized in Table 4.4.

According to the discussed lumped parameter model and parametric analysis, it can be found that the number of turns of coils can significantly increase the inductance; however, as the number of turns increased, the loss between the coil and the substrate will also increase, resulting in a rapid decrease in the Q value. Coil width should be as small as possible, while the increased line width will lead to an increase in the area above ground,

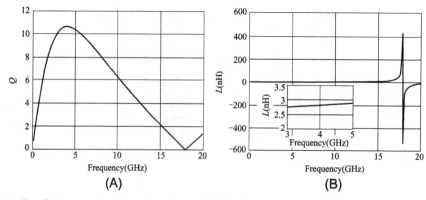

Fig. 4.11 Simulation results: (A) quality factor Q; (B) inductance.

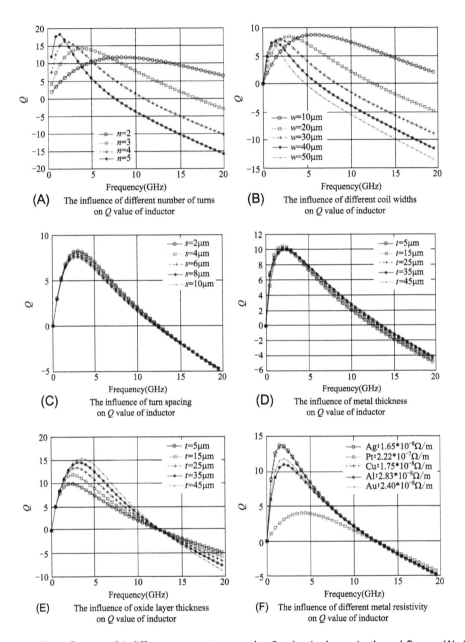

Fig. 4.12 The influence of 9 different parameters on the Q value is shown in the subfigures (A)–(I).
(Continued)

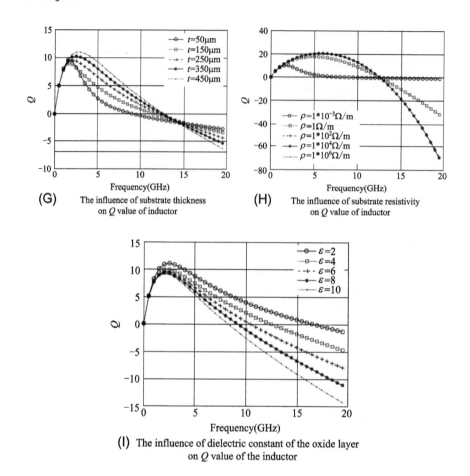

(G) The influence of substrate thickness on Q value of inductor

(H) The influence of substrate resistivity on Q value of inductor

(I) The influence of dielectric constant of the oxide layer on Q value of the inductor

Fig. 4.12, Cont'd

and therefore higher loss as well as lower Q value. At the same time, large line width can increase capacitance C_S of the parasitic feed-through capacitor between the upper and lower layers, resulting in a rapid decrease in resonant frequency as well as a decrease in the applicable interval of stable inductance. The inductive metal should be as thick as possible in the acceptable range. However, due to the presence of the skin effect, at high frequencies the improvement brought by increased thickness is not obvious. The spacing of the coil should be minimized, so that the mutual inductance between the straight segments can be enhanced, and the leakage loss associated with the coupling capacitor between the straight wire segments can be minimized, and thus a maximized Q value can be obtained. Materials with low resistivity should be used for the coil to effectively reduce the loss caused by skin effects. The insulation layer should be thickened as much as possible within the allowed range to increase the distance between substrates,

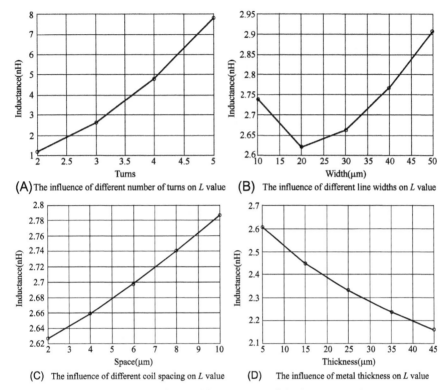

Fig. 4.13 The influence of coil turns, line width, spacing and thickness on inductance (A)–(D).

Table 4.4 Change of main performance parameters vs the number of turns and other parameters.

Inductor parameters	Q value change (the corresponding parameter increases)	L value change (the corresponding parameter increases)	Q peak frequency (the corresponding parameter increases)
Coil turns (n)	Decrease	Increase	Decrease
Coil width (w)	Decrease	Decrease first and then increase	Decrease
Coil thickness (t)	Increase to a certain value	Constant	Decrease to a certain value
Turn spacing (s)	Decrease	Increase	Almost constant
Coil material (Pt→Al→Au→Cu→Ag)	Decrease	Constant	Constant
Substrate thickness (t)	Constant	Constant	Constant
Insulation thickness (t)	Increase	Constant	Increase
Substrate resistivity (ρ)	Constant	Constant	Constant
Substrate dielectric constant (ε)	Decrease	Constant	Almost constant
Insulating layer dielectric constant (ε)	Decrease	Constant	Almost constant

such that the leakage loss associated with the coupling capacitors can be reduced. Material of small dielectric constant should be used as the substrate medium and the insulation medium. Thick medium substrate should be chosen to reduce the leakage loss associated with the medium substrate coupling capacitor; for a similar reason, a medium substrate with high resistivity is favored. However, when the resistivity of the medium substrate increases to a certain extent, the improvement of the Q value will be very small, so increasing the resistivity of the medium substrate is not always an effective way to increase the Q value.

Fig. 4.14 shows a suspended planar spiral inductor fabricated using an HR-Si TSV interposer process and released by performing local etching. As shown, the metal coil has been separated from the substrate, while the bending due to internal stress can also be observed. Fig. 4.15 shows the results of computation based on the modified model, the result of simulation analysis using HFSS, and the test results. In terms of quality factors, the modified model shows a better matching with the test results in the low-frequency range than the HFSS simulation. In the high-frequency range, HFSS works well, though there is still some difference in the test results. The reason lies in the failure in prediction of the proximity effect due to the bending of the inductor, while in the high-frequency range the proximity effect is actually the dominator among all the impedance factors having an impact on quality factors. At the same time, at high frequencies, the impact due to skin effect as well as surface roughness will be intensified, such that the path of surface conduction current can be further increased. In addition, at high frequencies, there is still leakage current inside the substrate, especially near the TSVs. However, compared to unsuspended inductors, the eddy current yielded by the inductor coil is smaller, which implies performing local etching on the surface of the substrate is a potential method to improve the quality factor of the inductor. In terms of inductance, the fitting result of

Fig. 4.14 HR-Si TSV interposer local floating inductor sample:(A) N2W40S10L400; (B) N2W40S10L600.

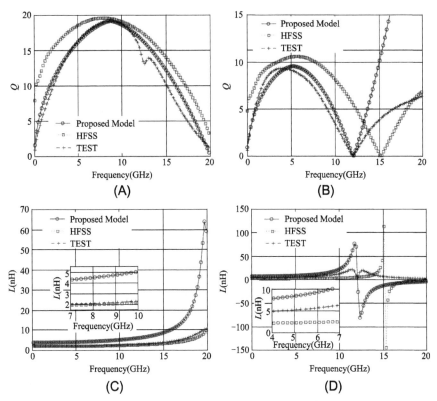

Fig. 4.15 Comparison of prediction results, simulation results, and actual measurement results: (A) Quality factor comparison of N2W0S10L400; (B) Quality factor comparison of N2W40S10L600; (C) Inductance comparison of N2W40S10L400; (D) Inductance comparison of N2W40S10L600.

N2W40S10L400 shows a significant difference, which may be due to the incomplete consideration of the impact on inductance associated with the eddy current. The fitting result is ideal for N2W40S10L600.

Subsequently, the method of optimization will be studied. The quality factor is required to more accurately estimate the effect due to surface roughness and the proximity effect in the high-frequency range. For inductance, the proximity effect and the equivalent inductance associated with surface roughness should be accurately introduced into the expression.

4.3 Research on 3D inductor based on HR-Si interposer

Fig. 4.16 shows the physical structure and geometrical parameters of an N-turn TSV-based toroidal inductor [32]; it is composed of four groups of conductor segments: TSVs at the inner ring, TSVs at the outer ring, RDLs on the top layer, and RDLs on the bottom layer.

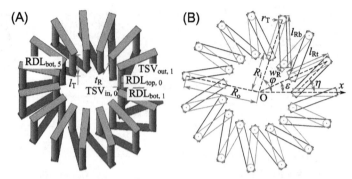

Fig. 4.16 Physical structure and geometrical parameters of toroidal inductor. (A) Perspective view. (B) Top view.

TSVs with length of l_T and radius of r_T are evenly distributed around the central axis with an angle $\varepsilon = 2\pi/N$; R_o and R_i denote the radii of the outer ring and inner ring, respectively. RDLs connect the TSVs at different rings to construct windings of the toroidal inductor; the width and thickness of RDLs are denoted as w_R and t_R; the length of the top RDL l_{Rt} and bottom RDL l_{Rb} can be calculated by Eq. (4.28). Each top RDL and adjacent bottom RDL in the counterclockwise direction forms an angle η, which can be calculated by Eq. (4.29). A rotation angle φ is defined as $\varphi = k\varepsilon$, $k = 0, 1, \cdots, N-1$, which is the angle formed by counterclockwise rotating of the x-axis about the toroid central axis to the line at which top RDLs are located. Accordingly, TSVs and RDLs are labeled in the counterclockwise direction as $TSV_{in,k}$, $TSV_{out,k}$, $RDL_{top,k}$ and $RDL_{bot,k}$, $k = 0, 1, \cdots, N-1$. Note that the TSV at input port ($TSV_{out,0}$) is removed, enabling the inductor to be connected to the other circuit.

$$l_{Rt} = R_o - R_i + 2r_T, \quad l_{Rb} = \sqrt{(R_o - R_i)^2 + 4R_o R_i \sin^2(\varepsilon/2)} + 2r_T \quad (4.28)$$

$$\eta = \arccos\left((R_o - R_i - 2R_o \sin^2(\varepsilon/2))/\sqrt{(R_o - R_i)^2 + 4R_o R_i \sin^2(\varepsilon/2)}\right) \quad (4.29)$$

The partial self-inductance of a single TSV and single RDL can be determined by Eq. (4.30) and Eq. (4.31), respectively, in which the internal inductance due to magnetic flux internal to conductors is incorporated. μ_0 is the vacuum permeability, $\mu_0 = 4\pi \times 10^{-7}$ H/m. Because $(2N-1)$ TSVs is connected in series in the toroidal inductor, the total self-inductance of TSVs can be expressed as Eq. (4.32). Similarly, the total self-inductance of RDLs can be expressed as Eq. (4.33), which is the series combination of N top RDLs and N bottom RDLs.

$$L_T(l_T, r_T) = \frac{\mu_0 l_T}{2\pi}\left[\sinh^{-1}\frac{l_T}{r_T} - \sqrt{1 + \left(\frac{r_T}{l_T}\right)^2} + \frac{r_T}{l_T} + \frac{1}{4}\right] \quad (4.30)$$

$$L_R(l_R, w_R, t_R) = \frac{\mu_0 l_R}{2\pi} \left[\ln\left(\frac{2l_R}{w_R + t_R}\right) + \frac{1}{2} + \left(\frac{w_R + t_R}{3l_R}\right) \right] \tag{4.31}$$

$$L_{T,\text{tot}} = (2N-1)L_T(l_T, r_T) \tag{4.32}$$

$$L_{R,\text{tot}} = N[L_R(l_{Rt}, w_R, t_R) + L_R(l_{Rb}, w_R, t_R)] \tag{4.33}$$

The partial mutual inductance of two TSVs, M_T, can be calculated by Eqs. (4.34) and (4.35) [16,17], where sig n_1 equals 1 if the currents in the two TSVs are in the same direction and equals -1 if the currents are in the opposite direction, l is the length of TSV, and d is the distance between the two TSVs. The expressions of d for two TSVs at inner ring ($d_{T,\text{in}}$), at outer ring ($d_{T,\text{out}}$), and at different rings ($d_{T,\text{dif}}$) are given in Eq. (4.36).

Taking the mutual inductance between $\text{TSV}_{\text{in},0}$ and other TSVs at the inner ring (i.e., $\text{TSV}_{\text{in},k}, k=1, 2, \cdots, N-1$) as a unit, the unit mutual inductance can be obtained by Eq. (4.37). Considering the rotational symmetry of the toroidal inductor, the total mutual inductance contributed by TSVs at the inner ring equals the unit mutual capacitance multiplied by the coefficient ($N/2$), as shown in Eq. (4.38). Similarly, the total mutual inductance of TSVs at the outer ring and at different rings can be calculated by Eqs. (4.39) and (4.40). The inductance arising from the removed TSV at the input port should be excluded; thus the coefficient in Eqs. (4.39) and (4.40) contains the term "-1".

$$M_1(l, d) = \frac{\mu_0}{2\pi} l \left[\operatorname{arcsinh} \frac{l}{d} - \sqrt{1 + \left(\frac{d}{l}\right)^2} + \frac{d}{l} \right] \tag{4.34}$$

$$M_T(l, d) = \text{sign}_1 \times M_1(l, d) \tag{4.35}$$

$$d_{T,\text{in}} = 2R_i \sin\left(\frac{\varphi}{2}\right), \, d_{T,\text{out}} = 2R_o \sin\left(\frac{\varphi}{2}\right), \, d_{T,\text{dif}} = \sqrt{R_o^2 + R_i^2 - 2R_o R_i \cos\varphi} \tag{4.36}$$

$$M_{T,\text{in,unit}} = \sum_{k=1}^{N-1} M_1(l_T, d_{T,\text{in}}) \tag{4.37}$$

$$M_{T,\text{in,tot}} = (N/2)M_{T,\text{in,unit}} = (N/2)\sum_{k=1}^{N-1} M_1(l_T, d_{T,\text{in}}) \tag{4.38}$$

$$M_{T,\text{out,tot}} = (N/2-1)\sum_{k=1}^{N-1} M_1(l_T, d_{T,\text{out}}) \tag{4.39}$$

$$M_{T,\text{dif,tot}} = (N-1)\sum_{k=0}^{N-1} [-M_1(l_T, d_{T,\text{dif}})] \tag{4.40}$$

The partial mutual inductance between nonparallel top RDLs and bottom RDLs can be obtained with formulas for mutual inductance of skewed and displaced conductors [29].

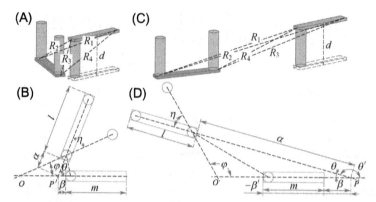

Fig. 4.17 Nonparallel RDLs from different layers. (A) Perspective view; (B) Top view of one example; (C) Perspective view; (D) Top view of another example

Taking $RDL_{top,0}$ and $RDL_{bot,1}$ in Fig. 4.16, for instance, Fig. 4.17A and B show the perspective view and top view of these two RDLs; the mutual inductance, $M_{R,np}$, can be calculated by Eq. (4.41), where $sign_2$ equals 1 if the currents in the two RDLs both flow toward or away from the intersection point P; otherwise $sign_2$ equals -1. $l=l_{Rb}$ is the length of bottom RDL, $m=l_{Rt}$ is the length of top RDL, d is the length of common perpendicular to the two RDLs, θ is the inclination angle, α and β are extension lengths to the intersection point P. R_1, R_2, R_3, and R_4 given in Eqs. (4.43)–(4.46) are the distances of endpoints of the two RDLs, Ω is defined in Eq. (4.47) for $d \neq 0$, and $\Omega = 0$ if $d = 0$. Substituting the expressions of α, β, and θ derived based on Fig. 4.17B into Eq. (4.42), the mutual inductance between $RDL_{top,0}$ and $RDL_{bot,1}$ can be expressed as Eq. (4.48).

$$M_{R,np}(l, m, d, \alpha, \beta, \theta) = sign_2 \times M_2(l, m, d, \alpha, \beta, \theta) \tag{4.41}$$

$$M_2(l, m, d, \alpha, \beta, \theta) = -\frac{\mu_0}{4\pi} \times \frac{\Omega d}{\tan\theta} + \frac{\mu_0 \cos\theta}{2\pi} \left[(\alpha + l) \operatorname{arctanh} \frac{m}{R_1 + R_2} + (\beta + m) \tanh^{-1} \frac{l}{R_1 + R_4} - \alpha \operatorname{arctanh} \frac{m}{R_3 + R_4} - \beta \operatorname{arctanh} \frac{l}{R_2 + R_3} \right] \tag{4.42}$$

$$R_1 = \sqrt{d^2 + (\alpha + l)^2 + (\beta + m)^2 - 2(\alpha + l)(\beta + m)\cos\theta} \tag{4.43}$$

$$R_2 = \sqrt{d^2 + (\alpha + l)^2 + \beta^2 - 2\beta(\alpha + l)\cos\theta} \tag{4.44}$$

$$R_3 = \sqrt{d^2 + \alpha^2 + \beta^2 - 2\alpha\beta\cos\theta} \tag{4.45}$$

$$R_4 = \sqrt{d^2 + \alpha^2 + (\beta + m)^2 - 2\alpha(\beta + m)\cos\theta} \tag{4.46}$$

$$\Omega = \tan^{-1}\frac{d^2\cos\theta + (\alpha+l)(\beta+m)\sin^2\theta}{dR_1\sin\theta} + \tan^{-1}\frac{d^2\cos\theta + \alpha\beta\sin^2\theta}{dR_3\sin\theta} - \tan^{-1}\frac{d^2\cos\theta + (\alpha+l)\beta\sin^2\theta}{dR_2\sin\theta} - \tan^{-1}\frac{d^2\cos\theta + \alpha(\beta+m)\sin^2\theta}{dR_4\sin\theta} \quad (4.47)$$

$$M_{R,pp,\text{dif}} = -M_2\left[\left(l_{Rb}, l_{Rt}, l_T + t_R, \frac{\sin\varphi}{\sin(\varphi+\eta)}R_i - r_T, \left(1 - \frac{\sin\eta}{\sin(\varphi+\eta)}\right)R_i - r_T, \varphi + \eta\right)\right] \quad (4.48)$$

The two nonparallel RDLs may be in several different relative positions. Fig. 4.17C and D depict another example ($RDL_{top,0}$ and $RDL_{bot,5}$ in Fig. 4.16). By introducing the definitions in Eq. (4.49), Eq. (4.50) can be derived according to Eqs. (4.43)–(4.47). Substituting Eq. (4.50) into Eq. (4.41), we can obtain Eq. (4.51), of which the left-hand side is exactly the mutual inductance between $RDL_{top,0}$ and $RDL_{bot,5}$ derived based on Eq. (4.41), and the righthand side is the same as that of Eq. (4.48) when substituting the expressions of α, β', and θ' in Fig. 4.17D into it. Thus Eq. (4.48) is also applicable to this example. It can be proven in a similar way that the mutual inductance between $RDL_{top,0}$ and each nonparallel bottom RDL can be expressed by the unified expression of Eq. (4.48). Therefore it is not necessary to calculate the mutual inductance between nonparallel RDLs case by case.

$$\begin{array}{l}\beta' = -(m+\beta), \theta' = \pi - \theta, \Omega' = \Omega(l, m, d, \alpha, \beta', \theta')\\ R'_j = R_j(l, m, d, \alpha, \beta', \theta'), j = 1, 2, 3, 4\end{array} \quad (4.49)$$

$$R_1 = R'_2, R_2 = R'_1, R_3 = R'_4, R_4 = R'_3, \Omega = \Omega' \quad (4.50)$$

$$M_2(l, m, d, \alpha, \beta, \theta) = -M_2(l, m, d, \alpha, \beta', \theta') \quad (4.51)$$

In addition, top RDLs and bottom RDLs may be parallel to each other. The prerequisites for parallel cases are $\varphi = \pi - \eta$ [case 1, shown in Fig. 4.18A and B] or $\varphi = 2\pi - \eta$ [case 2, shown in Fig. 4.18C and D]. The mutual inductance for parallel RDLs, $M_{R,p}$, can be written in terms of Eq. (4.34) as Eq. (4.53) [17], where d and s are center-to-center distance and endpoints offset of the two RDLs, respectively, and determination of the value of $sign_3$ follows the same rule as $sign_1$ in Eq. (4.35). Substituting d and s derived based on Fig. 4.18 into Eqs. (4.52) and (4.53), we could obtain Eqs. (4.54) and (4.55), which are the expressions of mutual inductance for the two parallel cases.

$$M_3(l, m, s, d) = 0.5[M_1(l + m + s, d) + M_1(s, d) - M_1(l + s, d) - M_1(m + s, d)] \quad (4.52)$$

$$M_{R,p}(l, m, s, d) = sign_3 \times M_3(l, m, s, d) \quad (4.53)$$

$$M_{R,p,\text{dif}1} = M_3\left[l_{Rb}, l_{Rt}, (1 - \cos\varphi)R_i - 2r_T, \sqrt{(R_i\sin\varphi)^2 + (l_T + t_R)^2}\right] \quad (4.54)$$

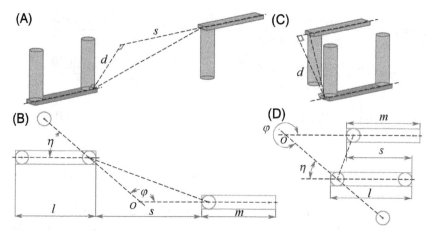

Fig. 4.18 The two parallel cases for RDLs from different layers. (A) Perspective view; (B) Top view of case 1; (C) Perspective view; (D) Top view of case 2.

$$M_{R,p,\,dif2} = -M_3\left[l_{Rb},\, l_{Rt},\, (1-\cos\varphi)R_i - l_{Rb},\, \sqrt{(R_i\sin\varphi)^2 + (l_T+t_R)^2}\right] \quad (4.55)$$

$$M_{R,\,dif,\,tot} = N\left(\sum_{\substack{0\leqslant k\leqslant N-1 \\ k\neq \frac{\pi-\eta}{\varepsilon},\,\frac{2\pi-\eta}{\varepsilon}}} M_{R,np,\,dif} + \sum_{\substack{0\leqslant k\leqslant N-1 \\ k=\frac{\pi-\eta}{\varepsilon}}} M_{R,p,\,dif1} + \sum_{\substack{0\leqslant k\leqslant N-1 \\ k=\frac{2\pi-\eta}{\varepsilon}}} M_{R,p,\,dif2}\right) \quad (4.56)$$

The mutual inductance between $RDL_{top,0}$ and all the bottom RDLs is considered as a unit, which times N is the total mutual inductance arising from RDLs from different layers (the expression is given in Eq. (4.56)). Similarly, the mutual inductance contributed by RDLs on the top layer and on the bottom layer can be obtained by Eqs. (4.57)–(4.59) and Eqs. (4.60)–(4.62), respectively.

$$M_{R,np,top} = M_2(l_{Rt},\, l_{Rt},\, 0,\, R_i - r_T,\, R_i - r_T,\, \varphi) \quad (4.57)$$

$$M_{R,p,top} = -M_3(l_{Rt},\, l_{Rt},\, 2(R_i - r_T),\, w_R/2) \quad (4.58)$$

$$M_{R,top,tot} = (N/2)\left(\sum_{1\leqslant k\leqslant N-1,\, k\neq\frac{\pi}{\varepsilon}} M_{R,np,top} + \sum_{1\leqslant k\leqslant N-1,\, k=\frac{\pi}{\varepsilon}} M_{R,p,top}\right) \quad (4.59)$$

$$M_{\text{R,np bot}} = M_2\left(l_{\text{Rb}}, l_{\text{Rb}}, 0, \frac{\cos(\varphi/2-\eta)}{\cos(\varphi/2)}R_i - r_{\text{T}}, \frac{\cos(\varphi/2+\eta)}{\cos(\varphi/2)}R_i - r_{\text{T}}, \varphi\right) \quad (4.60)$$

$$M_{\text{R,p,bot}} = -M_3[l_{\text{Rb}}, l_{\text{Rb}}, 2(R_i\cos\eta - r_{\text{T}}), 2R_i\sin\eta] \quad (4.61)$$

$$M_{\text{R,bot,tot}} = (N/2)\left(\sum_{1\leq k\leq N-1, k\neq\frac{\pi}{e}} M_{\text{R,np,bot}} + \sum_{1\leq k\leq N-1, k=\frac{\pi}{e}} M_{\text{R,p,bot}}\right) \quad (4.62)$$

The mutual inductance between TSVs and RDLs is zero because the currents inside them are orthogonal. The total inductance of the toroidal inductor is obtained by summing the self- and mutual inductance of all conductor segments as described in Eq. (4.63):

$$\begin{aligned}L_{\text{tot}} = &L_{\text{T,tot}} + L_{\text{R,tot}} + M_{\text{T,in tot}} + M_{\text{T,out tot}} + M_{\text{T, dif tot}} \\ &+ M_{\text{R,top,tot}} + M_{\text{R,bot,tot}} + M_{\text{R, dif tot}}\end{aligned} \quad (4.63)$$

The accuracy and efficiency of the inductance model are investigated in this section, and the results of modeling are compared with those from a standard simulator, Q3D extractor.

The toroidal inductor is modeled with an inner radius of 300 μm, outer radius of 600 μm, TSV length of 200 μm and radius of 10 μm, RDL width of 20 μm and thickness of 2 μm. Fig. 4.19 shows the calculated inductance and computation time of the two methods while sweeping the number of turns of the inductor. The inductance model demonstrates high accuracy with a percent error less than 6%. Furthermore, the computation time of the proposed model is about 0.03 s, much less than the time consumed by Q3D simulation; efficiency improvement is a considerable advantage of the model.

Fig. 4.20 compares the inductance value with various outer radius, inner radius, TSV height, and RDL width values. The results of modeling are in good agreement with those obtained by simulation, which also validates the accuracy of the model.

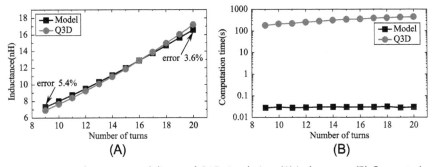

Fig. 4.19 Comparison between modeling and Q3D simulation: (A) Inductance; (B) Computation time.

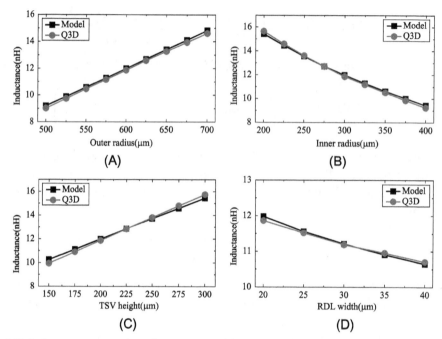

Fig. 4.20 Inductance comparison between modeling and simulation with various parameters: (A) Outer radius; (B) Inner radius; (C) TSV height; (D) RDL width.

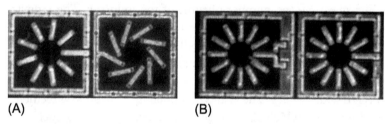

Fig. 4.21 TSV 3D inductor: (A) N9W70R30L370; (B)N12W70R30L370.

Fig. 4.21 shows the optical photo of the TSV 3D inductor. Fig. 4.22 shows the test results. Among them, the maximum measured inductance of the 12-turn structure inductance reaches 7.5 nH, and the error between the proposed theoretical model and the measured result is less than 6%.

Fig. 4.23 shows a more detailed comparison between the modeling result and the test result in terms of inductance. In the regions marked by gray rectangles, the modeling results stay the same after a slight drop, and deviate from the test results by no more than 10%. The inductances acquired by modeling are still constant out of the regions, while the test results show dramatic changes due to self-resonance between parasitic capacitor and real capacitor. While the impacts due to parasitic parameters associated with measurement are more

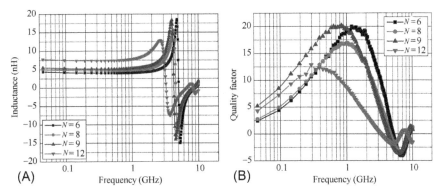

Fig. 4.22 TSV 3D inductor test results: (A) Inductance; (B) Q value.

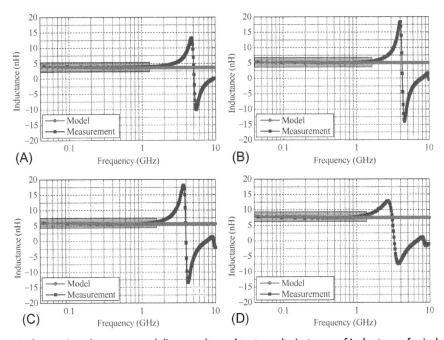

Fig. 4.23 Comparison between modeling results and test results in terms of inductance for inductors with different numbers of turns: (A) 6 turns; (B) 8 turns; (C) 9 turns; (D) 12 turns.

significant at high frequencies, the read-out value of inductance can greatly deviate from the real inductance, which should be a constant (i.e., the local minimum before self-resonance occurs). In this scenario, the inductance model is verified as valid.

Table 4.5 summarizes the research results on TSV 3D inductors published in recent years.

Table 4.5 Comparison of main performance parameters of HR-Si TSV interposer integrated inductors in recent years.

Literature	Date	Quality factor	Inductance	Resonant frequency (GHz)	Dimensions
CEA–Leti [11]	2019.5	16 at 0.4 GHz	6 nH	3	–
GIT [33]	2016.1	55 at 6.75 GHz	1.14 nH	21	$1530 \times 1030 \times 12\,\mu m^3$
GLA [29]	2018.1	22 at 24 GHz	0.83 nH	59	–
SILEX [31]	2013.10	14–134	1.38–11.91 nH	>10	–
NYU [14]	2015.5	9.1 at 2 GHz	4.52 nH	>15	$1.5 \times 1.5 \times 0.8\,mm^3$
This work	N9W70R30L370	20 at 0.8 GHz	5 nH	4.4	$1.85 \times 1.6 \times 0.28\,mm^3$
	N12W70R30L370	14 at 0.4 GHz	7.5 nH	3	$1.85 \times 1.6 \times 0.28\,mm^3$

4.4 Summary

In this chapter, a theoretical model of a traditional on-chip integrated planar spiral inductor and a method of parameter extraction are studied. Major factors impacting the performance of inductors are discussed. A design of a partial-etching-suspended planar spiral inductor on an HR-Si TSV interposer is proposed. Associated samples are fabricated and tested, and the results are analyzed. For the on-chip integrated inductor, a quality factor of 19.22 at 10 GHz is evaluated according to the test results. The average errors of Q value acquired by 3D electromagnetic simulation software HFSS (High Frequency Structure Simulator) and tests were 2.23 and 0.82, respectively. The accuracy and efficiency were verified.

In addition, a TSV-based 3D ring inductor model is also studied. The parasitic influence factors are analyzed. The accuracy of the model was verified by Q3D simulation analysis results as well as the test results. The maximum measured inductance of the 12-turn structure inductance reached 7.5 nH with a self-resonant frequency of 3 GHz.

References

[1] Jiadong C, Bin X. Overview of planar spiral inductance design and modeling[C]. In: International conference on microwave and millimeter wave technology; 2019.
[2] Sun Y, Cai H, Li J, et al. Design, fabrication and characterization of a novel TSV interposer integrated inductor for RF applications[C]. In: Electronic components and technology conference; 2018.
[3] Greenhouse HM. Design of planar rectangular microelectronic inductors. IEEE Trans Parts, Hybrids, Packaging 1974;(2):101–9.
[4] Wang X. Research on key technologies of front end of digital multimedia receving system[D]. In: Cheng du: University of Electronic Science and Technology of China, TN851; 2010 [in Chinese].
[5] Liu C. Research on silicon integrated inductors and CMOS RF integrated circuits[D]. Shang hai: Graduate University of the Chinese Academy of sciences(Shanghai Institute of Microsystems and Information Technology), TN431; 2002 [in Chinese].
[6] Chan RJ, Guo JC. Analysis and modeling of skin and proximity effects for millimeter-wave inductors design in nanoscale Si CMOS[C]. In: European microwave integrated circuit conference IEEE; 2014.
[7] Liu S, Zhu L, Allibert F, et al. Physical models of planar spiral inductor integrated on the high-resistivity and trap-rich silicon-on-insulator substrates[J]. IEEE Trans Electron Devices 2017;1–7.
[8] Bo H, Daimu W, et al. Poles–zeros analysis and broadband equivalent circuit for on-chip spiral inductors[J]. Int J Numer Modell Electron Networks Devices Fields 2016;29(3):446–57.
[9] Li Y, Sun S. Full-wave semi-analytical modeling of planar spiral inductors in layered media[J]. In: Progress in electromagnetics research-pier; 2014. p. 45–54.
[10] Pares G, Jean-Philippe M, Edouard D, et al. Highly compact RF transceiver module using high resistive silicon interposer with embedded inductors and heterogeneous dies integration[C]//2019 IEEE 69th electronic components and technology conference (ECTC). IEEE; 2019. p. 1279–86.
[11] Ho SW, Yoon SW, Zhou Q, et al. High RF performance TSV silicon carrier for high frequency application[C]. In: Electronic components and technology conference; 2008. p. 1946–52.
[12] Lazarus N, Bedair SS, Smith GL, et al. Origami inductors: rapid folding of 3-D coils on a laser cutter[J]. IEEE Electron Device Lett 2018;39(7):1046–9.
[13] Casu EA, Muller AA, Cavalieri M, et al. A reconfigurable inductor based on vanadium dioxide insulator-to-metal transition[J]. IEEE Microwave Wireless Compon Lett 2018;28(9):795–7.
[14] Kim B, Mondal S, Cho S, et al. A novel TSV inductor structure for RF applications[C]. In: Electronic components and technology conference; 2015. p. 236–9.

[15] Yook J, Kim D, Kim JC, et al. High-Q trenched spiral inductors and low-loss low pass filters using through silicon via processes[J]. Jpn J Appl Phys 2014;53(4).
[16] Pan J, Wang H, Qiu D, et al. Circuit modeling and structure optimization of integrated passive inductors[C]. In: International conference on electronic packaging technology; 2014. p. 1260–4.
[17] Yaya DD, Capraro S, Youssouf K, et al. Fabrication and properties of integrated magnetic inductors using an RLC model which takes into account eddy current[J]. Microsyst Technol 2016;23:3827–33.
[18] Wang C, Liao H, Xiong Y, et al. A physics-based equivalent-circuit model for on-Chip symmetric transformers with accurate substrate modeling[J]. IEEE Trans Microwave Theory Tech 2009;57(4):980–90.
[19] Yue CP, Ryu C, Lau J, et al. A physical model for planar spiral inductors on silicon[C]. In: International electron devices meeting; 1996. p. 155–8.
[20] Rotella F, Bhattacharya BK, Blaschke V, et al. A broad-band lumped element analytic model incorporating skin effect and substrate loss for inductors and inductor like components for silicon technology performance assessment and RFIC design[J]. IEEE Trans Electron Devices 2005;52(7):1429–41.
[21] Curran B, Ndip I, Guttowski S, et al. On the quantification and improvement of the models for surface roughness[C]. In: Workshop on signal propagation on interconnects; 2009. p. 1–4.
[22] Goni A, Pino JD, Gonzalez B, et al. An analytical model of electric substrate losses for planar spiral inductors on silicon[J]. IEEE Trans Electron Devices 2007;54(3):546–53.
[23] Feng Z, Bower CA, Carlson J, et al. High-Q solenoidal inductive elements[C]//2007 IEEE/MTT-S international microwave symposium. IEEE; 2007. p. 1905–8.
[24] Hall SH, Heck HL. Advanced signal integrity for high-speed digital designs. John Wiley; 2009.
[25] Yi M, Li S, Yu H, et al. Surface roughness modeling of substrate integrated waveguide in D-band[J]. IEEE Trans Microwave Theory Tech 2016;64(4):1209–16.
[26] Hu L, He S, Sun Y, et al. Design and process technology for high Q integrated inductor on interposer with TSV[C]. In: International conference on microwave and millimeter wave technology; 2018.
[27] Wojnowski M, Sommer G, Pressel K, et al. 3D eWLB—horizontal and vertical interconnects for integration of passive components[C]//2013 IEEE 63rd electronic components and technology conference. IEEE; 2013. p. 2121–5.
[28] Paul CR. Inductance: Loop and partial[M]. John Wiley & Sons; 2011.
[29] Eblabla A, Li X, Wallis DJ, et al. High-performance MMIC inductors for GaN-on-low-resistivity silicon for microwave applications[J]. IEEE Microwave Wireless Compon Lett 2018;28(2):99–101.
[30] Kuhn WB, Ibrahim NM. Approximate analytical modeling of current crowding effects in multi-turn spiral inductors[C]. In: International microwave symposium; 2000. p. 405–8.
[31] Ebefors T, Fredlund J, Perttu D, et al. The development and evaluation of RF TSV for 3D IPD applications[C]. In: IEEE International 3D Systems Integration Conference; 2013. p. 1–8.
[32] Feng Z, Bower CA, Carlson J, et al. High-Q Solenoidal Inductive Elements[C]// Microwave Symposium, 2007. IEEE/MTT-S International. IEEE; 2007.
[33] Thadesar PA, Bakir MS. Fabrication and characterization of polymer-enhanced TSVs, inductors, and antennas for mixed-signal silicon interposer platforms[J]. IEEE Trans Compon Packag Manuf Technol 2016;6(3):455–63.

CHAPTER 5

Verification of 2.5D/3D heterogeneous RF integration of HR-Si interposer

5.1 Introduction

Currently, whether in the field of advanced electronic information equipment or mobile communications equipment, system-in-package is a competitive integration scheme to realize miniaturized RF components such as transmitters/receivers. The state of the art of integration technology is mainly implemented by multichip modules (MCMs) based on high-performance substrates, such as microwave printed circuit boards low temperature cofired ceramics (LTCC), with RF/microwave microelectronic chips being mounted on one or both sides with an electromagnetic shielding structure or plastic molding, according to the application scenario. This method can make full use of the excellent performance of the matrix of high-performance chips fabricated on different substrates. However, there are issues, such as low precision, cofiring shrinkage mismatch, low thermal conductivity, and others [1], and the integration degree needs to be further improved. The HR-Si can achieve high-precision wiring, has the advantages of low thermal expansion coefficient mismatch with integrated microelectronic chips, and can provide high-quality passive components enabled by MEMS technology. It is a potential competitor for packaging substrates for 2.5/3D heterogeneous integrated RF systems. With the continuing requirements for miniaturization and integration in RF components, the high-resistivity Si will find a major opportunity. In 2015, the CEA-Leti (France) reported on HR-Si TSV interposer technology based on the RDL-first/TSV-last scheme, and that technology was used in a 3D RF integrated transceiver module [2]. The RF transceiver chip is mounted on the interposer by a flip-chip bonding process with a following EMC molding and then assembled to the PCB through solder ball to form a fully functional millimeter wave transceiver module. In 2019, CEA-Leti reported a 3D integrated TR component based on a HR-Si TSV interposer [3]. Compared with the equivalent module fabricated on an organic substrate, the final 2.5D heterogeneous integrated version showed a volume reduction of about 60%; the output power performance test results of the MMPA power amplifier were given, while the overall functional test results were not disclosed when published. In this chapter, we discuss our explorations in this field since 2014, with demonstration of a four-channel 2.5D heterogeneous integrated L band receiver, a 3D heterogeneous integrated 6–10 GHz channelized frequency conversion receiver based on a HR-Si TSV interposer.

5.2 Four-channel 2.5D heterogeneous integrated L-band receiver

In order to verify the feasibility of a HR-Si TSV interposer-based 2.5D/3D heterogeneous RF integration scheme, an L band receiver is utilized as a demonstrator and implemented [4,5]. Fig. 5.1 shows the schematic diagram of a four-channel L-band receiver, which is mainly composed of an low noise amplifier, splitters, attenuator, and phase shifting. Fig. 5.2 shows the S-parameters of the chips utilized in a 2.5D heterogeneous integrated L-band receiver. Fig. 5.3 shows the schematic diagram of a 2.5D heterogeneous RF integration scheme based on a HR-Si interposer. Here, the HR-Si TSV interposer has a layer of RDL of Cu on both sides to transmit the RF signal or logic IC signal. RF microelectronic chips or a driver IC chip are mounted on one side with wire-bonding or a flip-chip bonding process. The input/output of the demonstrator is implemented with an SMA connector through a CPW transmission line, TSV, bonding wire, and solder joint. In order to interact with the failure risk during in-process or work mode, the redundant TSV is designed as shown in Fig. 5.4; The failure of whichever one can cause about 0.2 dB loss in the L-band. Layout is finished for the HR-Si interposer as shown in Fig. 5.5, with a zoom-in with RF TSV. System-level simulation of a 2.5D heterogeneous integration L-band receiver is done with the extracted S-parameters, in combination

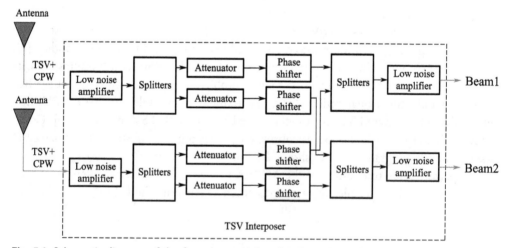

Fig. 5.1 Schematic diagram of the four-channel L band 2.5D integrated receiver.

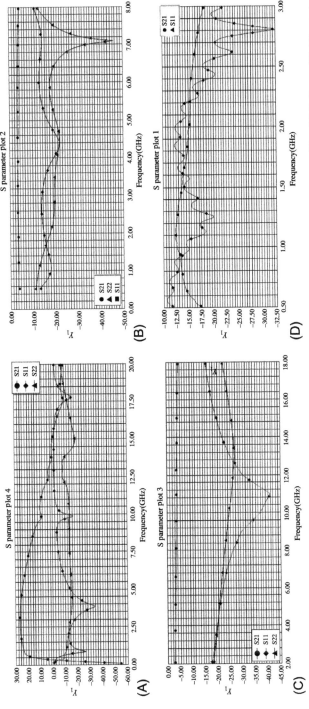

Fig. 5.2 S-parameters of the chips utilized in 2.5D heterogeneous integrated L-band receiver: (A) Low noise amplifier; (B) Splitters; (C) Attenuator; (D) Phase shifter.

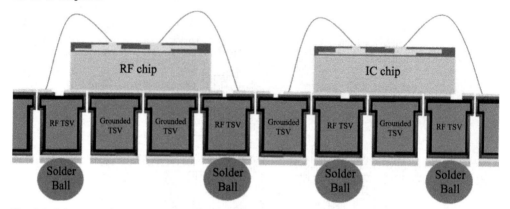

Fig. 5.3 Schematic diagram of 2.5D RF integrated scheme based on HR-Si TSV interposer.

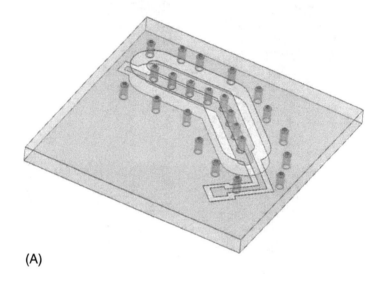

(A)

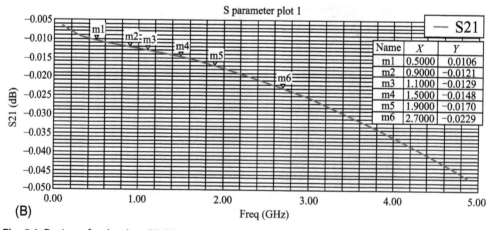

(B)

Fig. 5.4 Design of redundant RF TSVs.

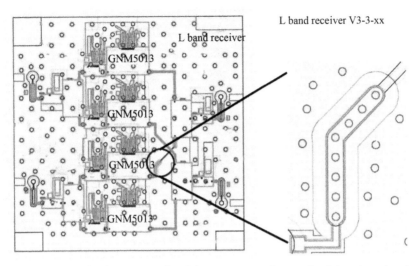

Fig. 5.5 Layout and partial enlargement of 2.5D heterogeneous integration L-band receiver based on HR-Si interposer.

with test S-parameters of RF chips; the results are illustrated in Fig. 5.6. The gain of the channel is around 25 dB.

Based on the developed process described in Chapters 2 and 3, a HR-Si interposer sample was fabricated as shown in Fig. 5.7. Fig. 5.8 illustrates the assembly flow of a four-channel 2.5D heterogeneous integrated L-band receiver. First, soldering was done with the HR-Si interposer sample and then it was assembled to a customized PCB package substrate mounted on a customized substrate; RF microelectronic dies were assembled

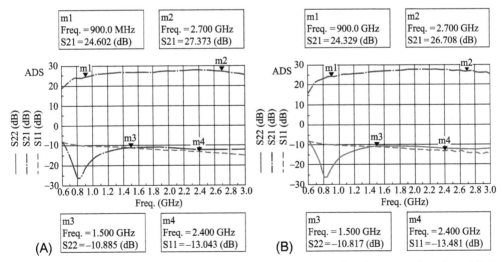

Fig. 5.6 Simulation results of 2.5D heterogeneous integration L-band receiver based on HR-Si interposer: (A) Beam 1; (B) Beam 2.

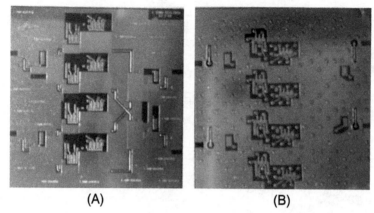

Fig. 5.7 Sample of HR-Si interposer for 2.5D heterogeneous integration L-band receiver based: (A) Surface of the small hole; (B) Surface of the big hole.

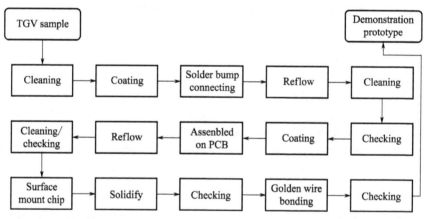

Fig. 5.8 The assembly flow of 2.5D heterogeneous integration L-band receiver based on HR-Si interposer.

on the HR-Si interposer using wire bonding technology. As the utilized chips were originally designed for mounting with wire-bonding, no flip-chip bonding was utilized in the demonstrator. Finally, the RF I/O interface was finished. Fig. 5.9 shows a photo of the HR-Si interposer after soldering and assembly of the RF chips. Fig. 5.10 shows the test curves for channel gain at a basic state; the input–output voltage standing wave ratio (VSWR) was tested and the function of phase shift and attenuation were experimentally verified, as summarized in Table 5.1. Comparing with the theoretical chain gain, the test value was about 2–3 dB lower. As the J6 – J5, J6 – J8, and J7 – J8 channels pass through two RF TSVs twice, J7 – J5 pass the RF TSV four times. The insertion loss is contributed by 13-mm CPW transmission lines and RF TSV. Factoring in the tested insertion loss of 0.2–0.3 dB/mm for the CPW line and 0.1–0.2 dB for RF TSV, this is basically consistent with the test results.

Fig. 5.9 Views during assembly of 2.5D heterogeneous integration L-band receiver based on HR-Si interposer.

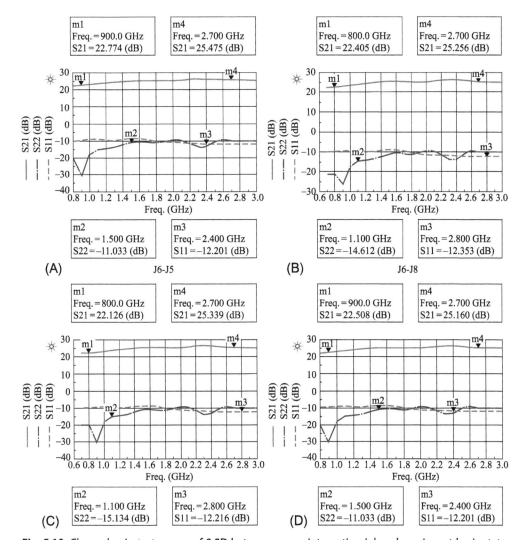

Fig. 5.10 Channel gain test curves of 2.5D heterogeneous integration L-band receiver at basic state.

Table 5.1 Four-channel L band 2.5D heterogeneous integrated receiver by HR-Si interposer.

Test result (0.8–2.7 GHz)	Gain (dB)	Input voltage standing wave	Output voltage standing wave	Phase shift	Attenuation
J6 – J5	22.7–25.5	2.2	1.9	Normal	Normal
J6 – J8	22.4–25.3	2.1	2	Normal	Normal
J7 – J5	22.1–25.4	2.2	1.9	Normal	Normal
J7 – J8	22.5–25.1	2.2	2	Normal	Normal

5.3 3D heterogeneous integrated channelized frequency conversion receiver based on HR-Si interposer

The channelized frequency conversion receiver uses a group of radio frequency filters to select a signal of a specific frequency that is received by the antenna, followed by signal mixing, filtering, amplification, modulation, and demodulation at the end. Generally, an attenuator and phase shifter are added to control the signal power and phase. Fig. 5.11 is the schematic diagram for the 6–10 GHz channelized frequency conversion receiver, which is utilized to verify the HR-Si interposer-based 3D heterogeneous RF integration scheme, which mainly includes a mixer, a local oscillator amplifier, an intermediate frequency amplifier, and a filter [6].

The radio-frequency filter is an indispensable component in the frequency conversion receiver, which plays the role of filtering and extracting specific frequency signals. In recent years, the research work has explored novel filter devices by using new structures, new materials, new processes, and new principles in this field. In 2017, the University of Virginia demonstrated a dual-frequency LC filter with a tunable high-quality factor using tunable resistors on a GeSi substrate; this work was funded by DARPA. The frequency and Q value of the two filters can be adjusted independently. The test results show the center frequency of the frequency band can be changed from 9.7 to 13.9 GHz, and the Q value of each band-pass filter can be adjusted from 20 to 70. A normalized dynamic

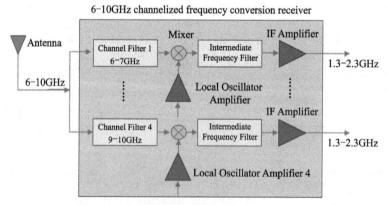

Fig. 5.11 Schematic diagram for 6–10 GHz channelized frequency conversion receiver.

change of 165–154.5 dB·Hz can be achieved with the overall center frequency changes between 9.7 and 13.9 GHz. The two passbands each have a bandwidth of 200 MHz, the isolation between the passbands is 22 dB, and the DC power consumption is 115–130 mW. In 2018, Georgia Institute of Technology reported on an interdigital filter based on a glass substrate, which is implemented with a high-precision redistribution layer (RDL) based on a semiadditive process (SAP). The test insertion loss of the filter is 2.6 dB, the rectangular coefficient corresponding to the attenuation point of 30 dB is 1.16, and the VSWR is better than 1.25. This is a popular way to make chip-level LC filters with CMOS technology, but there are some problems, such as more parasitic capacitance and inductance, and narrower frequency bands. In addition to the traditional distributed filter and LC filter composed of lumped elements, the substrate integrated waveguide filter is also an important developmental direction to realize integrated filters, with the characteristics of high application frequency and small insertion loss. In 2019, Dr. Xuanxuan Zhang from the Chinese Academy of Sciences designed and developed a K-band microstrip interdigital filter based on microelectromechanical systems (MEMS) technology.

The prototype of a channelized frequency conversion receiver based on an HR-Si interposer was proposed to work in the frequency range of 6–10 GHz, which was divided into four frequency channels, including 6–7 GHz, 7–8 GHz, 8–9 GHz, and 9–10 GHz. The intermediate frequency output is 1.3–2.3 GHz, the gain of the channel is required to be no less than 10 dB, and the power consumption is required to be less than 300 mW; isolation between channels is required to be no less than 40 dB. Among them, the channel filter uses a microstrip filter with an interdigital structure suitable for broadband filtering and has a compact structure, and therefore is implemented on the HR-Si interposer. Since the frequency of the intermediate frequency filter is 1.3–2.3 GHz, the frequency is low and the filter area of the microstrip line structure is large. Thus an on-chip LC filter is selected. Table 5.2 summarizes the main technical parameters of the selected

Table 5.2 The main technical parameters of the selected devices for the 3D heterogeneous integrated 6–10 GHz channelized frequency conversion receiver.

Device name	Device model	Frequency	Gain (dB)	Output P − 1 (dBm)	Power consumption
Mixer	WHDS044136-C70	LO&RF frequency range: 4.4–13.6 GHz IF frequency range: DC6.0 GHz	−7	2	−
Local oscillator amplifier	HMC-ALH482	2–22 GHz	16	14	180 mW
IF amplifier	GNA6102	10–2500 MHz	27	0	90 mW

devices for the 3D heterogeneous integrated 6–10 GHz channelized frequency conversion receiver.

5.3.1 HR-Si interposer integrated microstrip interdigital filter

Fig. 5.12 shows a schematic of the microstrip interdigital filter, which can be categorized into three types, namely the terminal open-circuit type, the terminal short-circuit type, and the capacitor-loaded type. The difference among them is whether the terminal admittance transmission lines W_0 and W_{n+1} are short-circuited or whether the microstrip resonators $W_1 - W_n$ in the middle are loaded with capacitors. Different from the microstrip interdigital filter of the open-terminal form, where each part is used as a resonant structure, the admittance transmission lines at both ends play the role of impedance matching when the interdigital filter is arranged in the terminal short-circuit form. The microstrip resonator in the middle is the main structure of the filter function, and the interdigital filter loaded with capacitance loads the end of each rod. It is believed that strengthening the end capacitance can shorten the length of the microstrip resonator.

The design of the microstrip interdigital filter can be implemented using a distributed parameter design method and lumped parameter design method. The former is mainly based on the transmission line theory to calculate the microwave components in the corresponding filter structure. The latter is divided into the parameter method and the network synthesis method. Between them, the network synthesis method is commonly used. It employs the network synthesis theory to normalize the design goals of the filter to obtain the structure parameters and value of the equivalent low-pass filter. Finally, this is followed by the frequency conversion method, calculating the self-capacitance and mutual capacitance of each microwave resonator with the proven formula, and using the capacitance network parameters to obtain the size parameters of the interdigital filter [7]. These traditional methods are relatively cumbersome. In order to improve the design

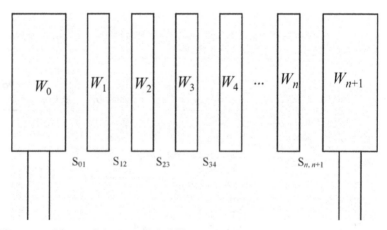

Fig. 5.12 The general form of the interdigital filter.

efficiency, this work adopts the design method of coupling coefficient [8]. The coupling coefficient design method first normalizes the design goal to the low-pass filter, obtaining the normalized parameters of the prototype filter, and then calculating the coupling coefficient and the length of the resonator.

According to previous studies, the substrate loss at high frequency is one of the major losses faced by integrated passive devices. On the basis of the aforementioned achievement in the HR-Si interposer, according to the application scenario requirements, an TSV integrated microstrip interdigital filter with a 3 dB bandwidth of 7.5–9.5 GHz, a ripple coefficient of 1 dB, and an attenuation greater than 20 dB at a frequency of 10.7 GHz is set as the design objective [9]. Here, it is taken as an example and the design process is expanded in detail. First, the low-pass prototype is mapped to the band-pass form. According to the design goal, a 5th-order Chebyshev low-pass prototype filter is selected. The corresponding prototype equivalent circuit is shown in Fig. 5.13. The normalized values of components have been listed in Table 5.3. Through Eqs. (5.1) and (5.2), the key adjacent resonator coupling coefficient $K_{i,i+1}$ and the filter terminal Q value can be calculated. According to the relationship between s/h and K, the corresponding resonator spacing and the width of the resonator are obtained. The length of the resonator is calculated according to Eq. (5.3), and the key dimensions of the filter are shown in Fig. 5.14. Similarly, the other three microstrip interdigital filters integrated on a high-R interposer can be designed and Table 5.4 summarizes the size parameters of the four filters.

$$K_{i,i+1} = \frac{FBW}{\sqrt{g_i g_{i+1}}} \quad (5.1)$$

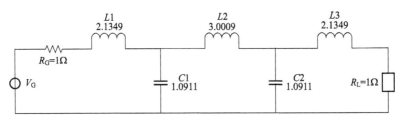

Fig. 5.13 Chebyshev T-type normalized equivalent low-pass circuit.

Table 5.3 UIR interdigital filter design parameters (frequency band: 7.5–9.5 GHz).

$g_0 = g_6$	$g_1 = g_5$	$g = g_4$	g_5
1.0000	2.11349	1.0911	3.0009
$K_{12} = K_{45}$	$K_{23} = K_{34}$	Q	
0.1542	0.1300	4.2500	

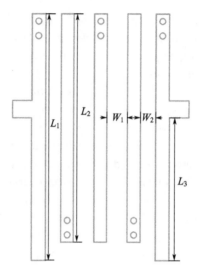

Fig. 5.14 Filter size diagram.

Table 5.4 Dimensional parameters of 6–10 GHz interdigital filter design.

Channel	L_1	L_2	L_3	W_1	W_2
5.5–7.5	4.6	4.2	2.5	0.2	0.13
6.5–8.5	4	3.7	2.3	0.26	0.18
7.5–9.5	3.44	3.2	2	0.28	0.21
8.5–10.5	3.13	2.9	1.9	0.33	0.24

$$Q = \frac{f_0}{BW} g_0 \tag{5.2}$$

$$L = \frac{1}{4} \times \frac{c}{f_0} \times \frac{1}{\sqrt{\varepsilon_r}} - 0.412h \frac{\varepsilon_e + 0.3}{\varepsilon_e - 0.258} \times \frac{w/h + 0.264}{w/h + 0.8} \tag{5.3}$$

Based on the design work aforementioned, verification and further optimization were conducted with HFSS. Fig. 5.15 shows the simulation results. The material parameters in the simulacrum are summarized in Table 5.5 and the simulation was performed in a frequency range up to 20 GHz. A 3 dB bandwidth of 2.64 GHz is obtained by the designed 7.5–9.5 GHz filter, the in-band ripple coefficient is 0.3 dB, the passband interpolation loss is 1.1118 dB, and the group time delay fluctuation is 0.052 ns. Fig. 5.16 shows the simulation results of the other three designed filters. Table 5.6 summarizes the simulation results of all four filters.

On the basis of the aforementioned achievement in the HR-Si TSV interposer, Fig. 5.17 shows the optical photo and X-ray inspection photo of one of the designed

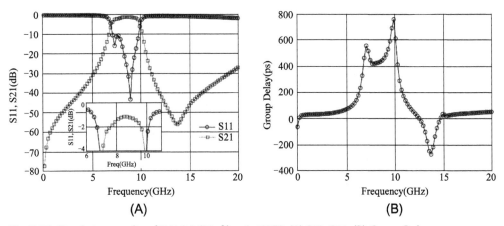

Fig. 5.15 Simulation results of 7.5–9.5 GHz filter in HFSS: (A) S11, S21; (B) Group Delay.

Table 5.5 Material parameters simulated in HFSS.

Silicon substrate		Copper
Bulk conductivity	Relative permittivity	Bulk conductivity
1/2500 S/m	11.9	5.8×10^7 S/m

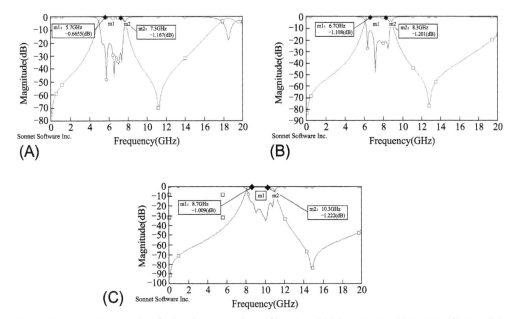

Fig. 5.16 Simulation results of other frequency band filters in HFSS: (A) 5.5–7.5; (B) 6.5–8.5; (C) 8.5–10.5.

Table 5.6 Simulation results of filter performance.

Frequency band (GHz)	In-band insertion loss	Out-of-band attenuation	Dimensions
5.5–7.5	<1.2 dB	<−20 dB at 4.3 GHz, <−20 dB at 8.3 GHz	3.2 mm × 5 mm
6.5–8.5	<1.2 dB	<−20 dB at 5.5 GHz, <−20 dB at 9.4 GHz	2.7 mm × 4.4 mm
7.5–9.5	<1.2 dB	<−20 dB at 6.4 GHz, <−20 dB at 10.7 GHz	2.8 mm × 3.7 mm
8.5–10.5	<1.2 dB	<−20 dB at 7.5 GHz, <−20 dB at 11.5 GHz	3 mm × 3.6 mm

HR-Si TSV interposer integrated microstrip interdigital filters. An Agilent vector network tester was used to test the samples. Compared with the theoretical results, the measured insertion loss was too large and the bandwidth was too small. To explore the reasons for this, the simulation was reconducted, considering the dimensional deviation due to process error; the results are shown in Fig. 5.18. It can be seen that the error of the thickness of the substrate has a greater impact on the bandwidth and the width error effects on the bandwidth. The large loss should not be ascribed to the dimension error.

To better understand the reason for the large insertion loss, the equivalent circuit is established as shown in Fig. 5.19, where the open circuit part and the short circuit part are represented as series resonators (C_{Si}, L_{Si}) and parallel resonators (C_{Pi}, L_{Pi}), respectively [10]. The main loss of the filter is self-loss R_S and substrate loss R_L. The influence of the coupling between the resonators is represented by the capacitor ($C_{i,\,i+1}$). The loss

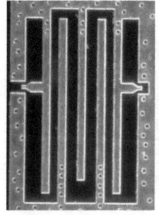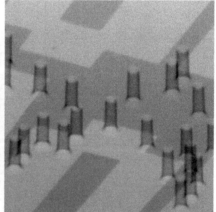

Fig. 5.17 Physical image and X-ray image.

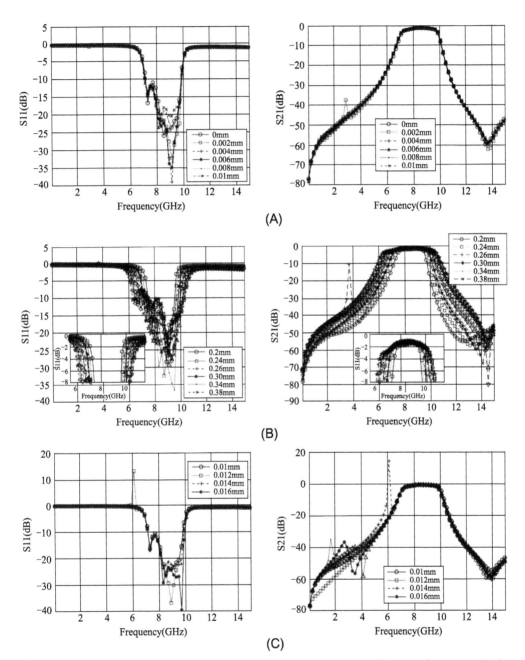

Fig. 5.18 Effects of the dimensional deviation due to process error on filter's performance: (A) the influence of line width error; (B) the influence of substrate thickness; (C) the influence of the thickness of metal layer.

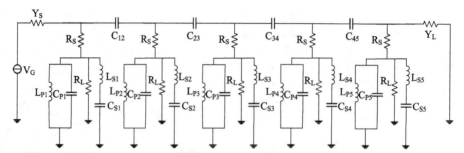

Fig. 5.19 Equivalent circuit model of interdigital filter.

caused by the substrate is included in the R_L, and the loss caused by the surface roughness is also included in the R_S. The main reason for the former is that the surface roughness of the electroplated metal does not represent an ideal smooth surface, resulting in additional equivalent resistance. At the same time, impurities may be mixed in during the electroplating process, resulting in unsatisfactory conductivity. The low bandwidth is mainly due to the classical methods used in the design process. In fact, most of the design methods are not completely based on the interaction of actual electromagnetic physics. The design formula is based on empirical data instead of actual physical field effects. The data are derived from the strict theoretical formula obtained by the analysis, so there will be certain deviations.

Based on the aforementioned analysis, the passband loss may be contributed by the electroplated Cu layer. Here, annealing and plasma treatment are carried out to improve the RF performance. Fig. 5.20 shows the insertion loss changes of the filter before and

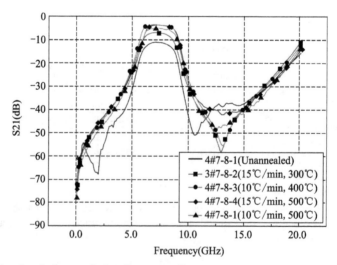

Fig. 5.20 Insertion loss before and after filter annealing.

Table 5.7 Band interpolation loss before and after annealing.

Filter channel	Loss (dB)
4#7-8-1 (Unannealed)	−11.2
4#7-8-1 (10°C/min, 500°C)	−5
3#7-8-3 (10°C/min, 400°C)	−3.7
3#7-8-2 (15°C/min, 300°C)	−7
4#7-8-4 (15°C/min, 500°C)	−3.9

after annealing. The annealing temperatures are set to 300°C, 400°C, and 500°C. The specific test results are shown in Table 5.7. It can be found that the annealing helps to reduce the insertion loss and the insertion loss of the filter is reduced from 11.2 dB to 3.7 dB after annealing at 400°C. Fig. 5.21 shows the change in insertion loss of the filter before and after plasma treatment. The test results are shown in Table 5.8. It can be concluded that the insertion loss is roughly reduced by 6–7 dB, and the minimum in-band insertion loss can reach 2.19 dB, which is close to the simulation result. Table 5.9 summarizes the measurement results of microstrip interdigital filter samples published in

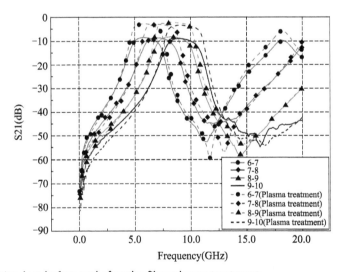

Fig. 5.21 Insertion loss before and after the filter plasma treatment.

Table 5.8 Band interpolation loss before and after plasma treatment.

Filter	Untreated (dB)	Plasma treatment (dB)
3#6-7-1	−8.57	−2.85
3#7-8-1	−8.70	−3.05
3#8-9-1	−8.27	−2.19
3#9-10-1	−9.01	−3.35

Table 5.9 Filter test performance published in recent years.

Publication	Time	Operating frequency/ bandwidth (GHz)	Insertion loss (dB)	Stopband suppression	Dimension
ONSEMI [11]	2019.4	3.3–4.2/0.9	⩽2.2	⩾30 dB at 2.7 and 5.18 GHz	1.6 mm × 0.8 mm
CETC ISA [12]	2018.11	3.3–3.6/0.3.	⩾2.8	⩾15 dB	1.5 mm × 0.5 mm
ONSEMI [13]	2019.9	5–20/5.2–9	3.6–4.6	⩾40 dB	2 mm × 0.25 mm
KETI [14]	2019.10	28	⩾1.1	–	0.492 mm × 0.88 mm
CETC 58 [15]	2018.8	2.45/0.3	⩾7	⩾15 dB	2.5 mm × 3.0 mm
This work		6.5/2.09	⩾2.1	⩾40 dB	2.8 mm × 3.7 mm
		7.5/2.08	⩾2.17	⩾40 dB	2.7 mm × 4.4 mm
		8.5/2.33	⩾2.2	⩾40 dB	2.8 mm × 3.7 mm
		9.5/2	⩾2.05	⩾40 dB	3 mm × 3.6 mm

recent years. The filter presented here is comparable to the filters reported by other research groups in terms of band interpolation loss, stopband suppression, and size.

5.3.2 Design, fabrication, and test of HR-Si interposer

High isolation is required between RF signal transmission channels in the HR-Si interposer, as the electromagnetic coupling among channels deteriorates further and is difficult to treat due to the increase in integration level, which has a decisive influence on the whole performance. A grounded TSV array is added around the RF signal transmission channel. Figs. 5.22 and 5.23 compare the isolation when with or without presence of a grounded TSV array for the CPW lines. Figs. 5.24 and 5.25 compare the isolation when

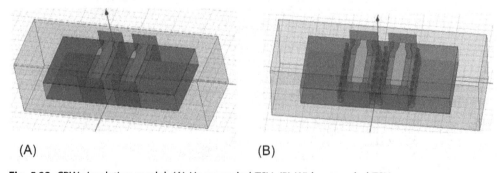

Fig. 5.22 CPW simulation model: (A) Ungrounded TSV; (B) With grounded TSV.

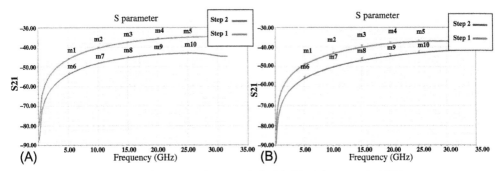

Fig. 5.23 CPW simulation isolation: (A) Ungrounded TSV; (B) With grounded TSV.

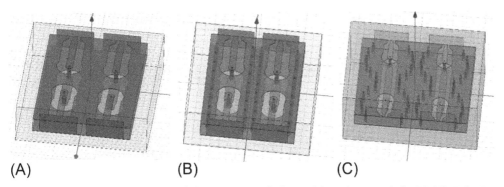

Fig. 5.24 CPW+TSV simulation model: (A) Ungrounded TSV; (B) With grounded TSV; (C) With ring grounded TSV array.

the parallel RF TSV linked CPW lines with the presence or absence of a grounded TSV array. The simulation results are summarized in Table 5.10. For parallel CPW lines, S31 with grounded TSV is 2–3 dB smaller than that without a grounded TSV array. In addition, when the distance is reduced by 100 μm, the increase in S31 with grounded TSV is smaller than that without grounded TSV. For the RF TSV linked CPW lines, the S31 curve with TSV grounding is much smoother than without TSV grounding and the value is tens of dB smaller; the S31 value without grounded TSV suddenly increases to −15 dB at 16 GHz. However, this situation would not appear with grounded TSV. Furthermore, it can be seen that when the distance is reduced, the increase in S31 with grounded TSV is smaller than that without grounded TSV. Therefore the impact of the crosstalk between the CPW lines or RF TSV linked CPW lines will be reduced when the TSV is grounded. At the same time, the isolation of the RF TSV linked CPW with a ring-grounded TSV array is also simulated, and the isolation is greater than 65 dB, which is better than the isolation effect of the in-line grounded TSV array.

Based on the isolation analysis, a serial TSV grounding scheme was implemented with the 3D heterogeneous integrated 6–10 GHz channelized frequency conversion receiver.

144 TSV 3D RF integration

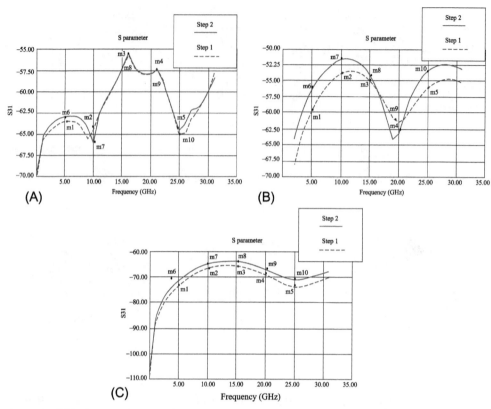

Fig. 5.25 CPW+TSV simulation isolation: (A) Ungrounded TSV; (B) With grounded TSV; (C) With ring grounded TSV array.

Table 5.10 Simulation results of transmission structure isolation.

Test structures		Isolation in different frequency (dB)					The change of isolation within 100 μm
		5 GHz	10 GHz	15 GHz	20 GHz	25 GHz	
CPW	Ungrounded TSV	53.8	48.3	45.5	43.9	43.2	−8
	Grounded TSV	55.9	50.2	46.9	44.6	43.1	−6
CPW +TSV	Ungrounded TSV	55.0	59.9	16.5	27.2	62.8	−3
	Grounded TSV	59.6	53.8	54.9	61.5	56	−2.5
	Circular grounded TSV array	73.3	66.6	65.7	69.0	73.8	−1.5

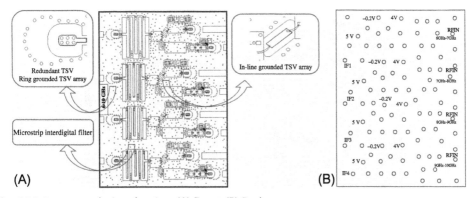

Fig. 5.26 Prototype design drawing: (A) Front; (B) Back.

Fig. 5.26 shows the layout for the HR-Si interposer. Note that a ring grounded TSV array is placed around the RF TSV to improve the isolation. In order to improve the capacity for antifailure risk, the input electrical signals from outside are first transmitted from the back of the HR-Si interposer to the front through the coaxial-like redundant TSV structure, and the signals in the 6–10 GHz frequency band are selected by the interdigital filter. The mix is followed with the signal generated by the local oscillator, and then transmitted to the intermediate frequency filter through the RDL, and the 1.3–2.3 GHz frequency band signal is obtained and sent out. After the intermediate frequency is amplified, it is transmitted out from the HR-Si interposer through the coaxial-like redundant TSV structure. The surface area of the sample is 15.94 mm × 6.6 mm. The RF signal loss in the transmission link is mainly contributed by two TSV transitions and RDL (the estimated link length is about 3.5 mm); the estimated link loss is 2 dB, and the overall (without filter) gain is 18 dB. The HR-Si interposer was fabricated with the developed process in our lab and the test results of the HR-Si interposer integrated filter are shown in Fig. 5.27. Passband loss is about 2 dB.

5.3.3 3D heterogeneous integrated assembly and test

After completing the passivation and posttreatment, the HR-Si TSV interposer is assembled and tested. The assembly steps include: the HR-Si TSV interposer is cleaned and soldering is conducted on the sides with a large hole. Secondly, the HR-Si TSV interposer is mounted onto a customized PCB board with solder array, reflowed, and cleaned again. The RF microelectronic die is assembled on the HR-Si TSV interposer. Finally, the RF chip is electrically connected with the HR-Si interposer by gold wire bonding. The assembled 3D heterogeneous channelized frequency conversion receiver is shown in Fig. 5.28; its size is 20.5 mm × 14.8 mm × 1.6 mm.

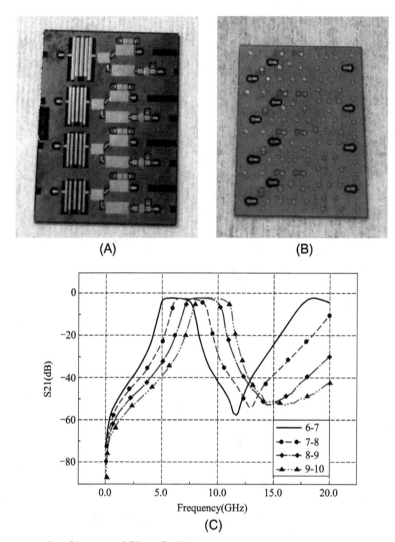

Fig. 5.27 Test results of integrated filter of HR-Si interposer.

The 3D heterogeneous integrated channelized frequency conversion receiver on the HR-Si interposer has four channels. The working frequency, function, gain, and isolation of the beam components of each channel are tested; the electrical performance test block diagram and test platform are shown in Fig. 5.29. The specific test process is as follows:

(1) Set the initial frequency of frequency source 1–6, 7, 8, 9 GHz, with an input power of −20 dBm, and connect to 6–7, 7–8, 8–9, and 9–10 ports, respectively;

(2) Set the frequency of the frequency source 2–4.7, 5.7, 6.7, 7.7 GHz, with an input power of −2 dBm, connect to LO1, LO2, LO3, and LO4 ports, respectively;

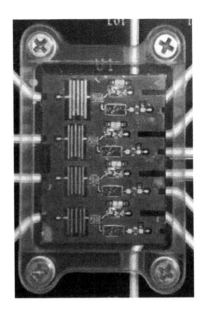

(A) Front view

(B) Side view

Fig. 5.28 Channelized frequency conversion receiver on the test fixture.

(3) Set the frequency of the spectrum analyzer to 1.3–2.3 GHz and connect to IF1, IF2, IF3, IF4 ports, respectively;
(4) The power supply is set in two working states: output voltage +4 V, current limit 50 mA, and voltage +5 V, current limit 20 mA;
(5) Change the frequency of frequency source 1, stepped by 200 MHz, until increased by 1 GHz, record the output power a of IF1, IF2, IF3, IF4 ports, and calculate the channel gain $= (a - 20)$ dB;
(6) Connect frequency source 1 to channels 6–7, and record the power output b of IF2 port, and calculate channel isolation $= (b - 20)$ dB.

Record the voltage V and current I when the previously mentioned prototype is working, and calculate the power consumption $= VI$. Through this test, four channels of the prototype receiving and down-conversion functions are verified.

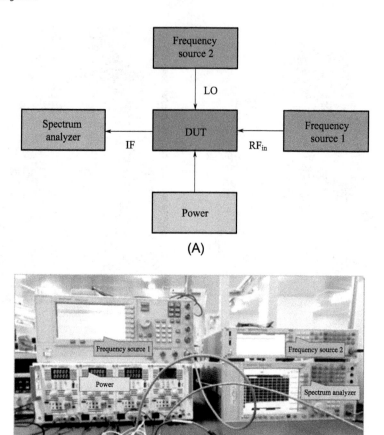

Fig. 5.29 Sample electromechanical performance test: (A) Electrical performance test block diagram; (B) Test bench.

The test results of each channel gain are shown in Tables 5.11–5.14. Among them, the line losses of the four channels of 6–7, 7–10, 8–9, and 9–10 GHz are 3, 4, 4.7, and 5.5 dB, respectively. It can be seen that the calculated gains of the four channels are ⩾13.5, ⩾13.3, ⩾16.8, and ⩾10.1 dB, respectively, which meet the gain target. Meanwhile, we measured the isolation between the 6–7 and 7–8 channels, and the result is >50 dB to meet the isolation target, as shown in Table 5.15. In the two working states of the power supply, the currents and voltages of channel 6–7 are 15.2 mA/5 V and 48.6 mA/4 V; the currents and voltages of channel 7–8 are 15.3 mA/5 V and 47.1 mA/4 V; the currents and voltages of channel 8–9 are 15.3 mA/5 V and 47.1 mA/4 V; the currents and voltages of channel 9–10 are 15.3 mA/5 V and

Verification of 2.5D/3D heterogeneous RF integration of HR-Si Interposer 149

Table 5.11 Prototype channel 6-7 test results.

RF$_{in}$ input frequency (GHz)	IF output frequency (GHz)	IF output frequency (dBm)	Gain (dB) (IF − RF-line loss)
6	1.3	−4.75	18.25
6.1	1.4	−5.8	17.2
6.2	1.5	−6	17
6.3	1.6	−7	16
6.4	1.7	−8	15
6.5	1.8	−8.5	14.5
6.6	1.9	−9.5	13.5
6.7	2	−9.4	13.6
6.8	2.1	−9	14
6.9	2.2	−8.4	14.6
7	2.3	−6.5	16.5

Table 5.12 Prototype channel 7-8 test results.

RF$_{in}$ input frequency (GHz)	IF output frequency (GHz)	IF output frequency (dBm)	Gain (dB) (IF − RF-line loss)
7	1.3	−6.35	17.65
7.1	1.4	−6.65	17.35
7.2	1.5	−8	16
7.3	1.6	−8.7	15.3
7.4	1.7	−8.5	15.5
7.5	1.8	−9.6	14.4
7.6	1.9	−10.7	13.3
7.7	2	−9.4	14.6
7.8	2.1	7.85	16.15
7.9	2.2	−7.9	16.1
8	2.3	−5.35	18.65

Table 5.13 Prototype channel 8-9 test results.

RF$_{in}$ input frequency (GHz)	IF output frequency (GHz)	IF output frequency (dBm)	Gain (dB) (IF − RF-line loss)
9	1.3	−7.4	17.3
8.9	1.4	−7.3	17.4
8.8	1.5	−7.9	16.8
8.7	1.6	−8	16.7
8.6	1.7	−7.2	17.5
8.5	1.8	−6.7	18
8.4	1.9	−6.7	18
8.3	2	−6	18.7
8.2	2.1	−5.9	18.8
8.1	2.2	−7	17.7
8	2.3	−7.5	17.2

Table 5.14 Prototype channel 9-10 test results.

RF$_{in}$ input frequency (GHz)	IF output frequency (GHz)	IF output frequency (dBm)	Gain (dB) (IF − RF-line loss)
10	1.3	−14.7	10.8
9.9	1.4	−15.2	10.3
9.8	1.5	−15	10.5
9.7	1.6	−15.1	10.4
9.6	1.7	−15.3	10.2
9.5	1.8	−15.4	10.1
9.4	1.9	−15	10.5
9.3	2	−14	11.5
9.2	2.1	−13.2	12.3
9.1	2.2	−13.2	12.3
9	2.3	−13.6	11.9

Table 5.15 Channel isolation.

6-7 channel RF$_{in}$ (GHz)	7-8 channel IF output frequency (GHz)	7-8 channel IF output frequency (dBm)	Isolation (dB) (IF − RF-line loss)
7	1.3	−74.65	51.65
7.1	1.4	−74.95	51.95
7.2	1.5	−76.3	53.3
7.3	1.6	−77	54
7.4	1.7	−76.8	53.8
7.5	1.8	−77.9	54.9
7.6	1.9	−79	56
7.7	2	−77.7	54.7
7.8	2.1	−76.15	53.15
7.9	2.2	−76.2	53.2
8	2.3	−73.65	50.65

47.1 mA/4 V. Among them, the maximum current is 15.2 mA/5 V, 48.6 mA/4 V, power consumption = 15.2 × 5 + 48.6 × 4 = 270.4 mW, which meets the power consumption target. The overall prototype realizes the function of frequency conversion from 6–10 GHz to 1.3–2.3 GHz, and ensures the goal of high gain, low loss, and high isolation of the channel.

5.4 Conclusions

This chapter examines the results of our team's explorations in the application of three-dimensional heterogeneous RF integration of an HR-Si. A 2.5D heterogeneous integrated four-channel L-band receiver based on the HR-Si interposer is demonstrated;

it has a tested channel gain of about 25 dB, and the redundant design of RF TSVs is verified. The feasibility is showcased. A 3D integrated 6–10 GHz channelized frequency conversion receiver based on the HR-Si interposer is displayed, which realized a channel gain ⩾10 dB, channel isolation >50 dB, transmission loss <5.5 dB, power consumption <300 mW, and frequency conversion function. The feasibility of the redundancy design and the coaxial design method is verified and the foundation is laid for the subsequent 3D stacked radio-frequency integration based on the HR-Si with higher frequency bands and higher wiring density.

References

[1] Lamy Y, Dussopt L, et al. A compact 3D silicon interposer package with integrated antenna for 60GHz wireless applications[C]. In: 2013 IEEE international 3D systems integration conference (3DIC), 264; 2013. p. 1–6.
[2] El Bouayadi O, Dussopt L, Lamy Y, et al. Silicon interposer: A versatile platform towards full-3D integration of wireless systems at millimeter-wave frequencies[C]. In: 2015 IEEE 65th electronic components and technology conference (ECTC). IEEE; 2015. p. 973–80.
[3] Pares G, Jean-Philippe M, Edouard D, et al. Highly compact RF transceiver module using high resistive silicon interposer with embedded inductors and heterogeneous dies integration[C]//2019 IEEE 69th electronic components and technology conference (ECTC). IEEE; 2019. p. 1279–86.
[4] Ma S. A 2.5D integrated L band receiver based on high-R Si interposer[C] [IEEE international conference on integrated circuits, technology and applications, 21–23 November]; 2018. Beijing,China.
[5] Yan J. Fundamental research on process and application of 3D RF integrated low loss TSV interposer [D]. Xiamen: Xiamen University; 2018 [in Chinese].
[6] Wang M. Design and application verification of 3D radio frequency interconnection based on TSV[D]. Xiamen: Xiamen University; 2020 [in Chinese].
[7] Xu J. The miniaturizaition Research of Interdigital Microstrip Bandpass Filter[D]. Tianjin: Hebei University of Technology (in Chinese); 2010.
[8] Matthaei GL, Young L, Jones EMT. Microwave filters, impedance matching networks and coupling structures[M]. New York: Mc Graw-Hill; 1964.
[9] Sun Y, Jin Y, Ma S, et al. Design, fabrication and measurement of TSV interposer integrated X-band microstrip filter[C]. In: International conference on electronic packaging technology (ICEPT); 2019.
[10] Fei W. Research on miniaturized microstrip antenna[D]. Nanjing: Nanjing University of Posts and Telecommunications; 2019 [in Chinese].
[11] Shin KR, Arendell J, Eilert K, et al. Compact 5G n77 band pass filter with through silicon via (TSV) IPD technology[C]. In: Wireless and microwave technology conference; 2019.
[12] [116] Ban Y, Liu J, Li W, Wang Z, et al. A miniaturized bandpass filter design and verification with de-embedding technology in SiP solutions[C]. In: International conference on integrated circuits and microsystems; 2018.
[13] Kim D, Min B, Yook J, et al. Compact mm-wave bandpass filters using silicon integrated passive device technology[J]. IEEE Microwave Wireless Compon Lett 2019;29(10):638–40.
[14] Yook J, Kim D, Park B, et al. A compact 28 GHz RF front-end module using IPDs and wafer-level metal fan-out packaging[C]. In: European microwave conference; 2019.
[15] Mao C, Zhu Y, Li Z, et al. Design of LC bandpass filters based on silicon-based IPD technology[C]. In: International conference on electronic packaging technology; 2018.

CHAPTER 6

HR-Si interposer embedded microchannel

6.1 Introduction

Transmitter/receiver components are indispensable functional units in advanced electronic information equipment, such as electronic countermeasures, radars, and communications, and their development toward miniaturization, integration, and high-performance is an important driving force for the development of TSV 3D RF heterogeneous integration technology. GaN material has the characteristics of wide bandgap, high thermal conductivity, high breakdown voltage, and high carrier saturation speed. A GaN HEMT device is capable of high-power and high-frequency applications, which are the enabling technology to support the development of high-performance TR components, and have broad application prospects. However, the application of high-power GaN HEMT devices also brings thorny heat dissipation problems. The power of a single GaN power amplifier device can reach up to 300 W, and the local heat flux density will exceed $1000\,\text{W}/\text{cm}^2$. If the heat cannot be dissipated in time, this will affect the performance of the chip or even cause chip failure [1–9].

In view of the heat dissipation problem of high-performance TR components, the current heat dissipation solutions can be divided into external fluid or microfluidics cooling, package embedded microfluidics cooling, and embedded microfluidics cooling, among other solutions.

For a typical external fluid or microfluidics cooling scenario, high-power GaN HEMT chips are usually mounted on high thermal conductivity carriers such as molybdenum copper carriers [10] or diamond carriers [11–15], and then loaded on a cold plate for heat dissipation where fluid or microfluidics cooling is applied. Fig. 6.1 shows a representative diagram for high-power GaN cooling reported by the 38th Research Institute of China Electronics Technology Group Corporation in 2020 [16]. The bare GaN chip is soldered on the Mo80Cu20 substrate by gold-tin solder, tin-lead solder is used to weld on the T/R component shell, and then the T/R component shell is assembled on the cold plate by screws; thermal interface material is used between the two to reduce the thermal resistance of the interface and 65% ethylene glycol solution is used to take away heat. In this scheme, the thermal resistance of the heat conduction path is large between the chip (heat source) and the heat sink, which limits the heat dissipation capacity of the terminal cold plate.

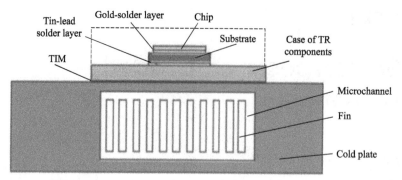

Fig. 6.1 Diagram of high-power GaN device based TR module with external fluid cooling.

The main drawback of external fluid or microfluidics cooling is due to the long heat transfer path from the heat source to the terminal cold plate, so it is a natural choice to extend the cooling medium directly into the package body, which means that the macroscale flow channel turns into a microchannel to dissipate heat. In the 1980s, Tuckerman and Pease of Stanford University [17–19] proposed the concept of microchannel heat dissipation. They used a MEMS process to fabricate parallel microchannel heat sinks on a silicon substrate. The channel width was 50 μm and the height was about 300 μm. The theoretical calculation and analysis showed that the temperature of the heat source assigned on the surface was reduced to about 71 °C when the input power density reached 790 W/cm^2, and the total thermal resistance in the path that started from the heat source to microfluidics was 0.09 °C/W, revealing a potential technological advantage in heat dissipation. Peng et al. [20] manufactured a variety of parallel microchannels with rectangular cross sections of different sizes and found that the hydraulic diameters and aspect ratios had a great influence on the heat transfer capacity, where the hydraulic diameter is defined as four times the ratio of the flow section area to the perimeter.

Owhaib et al. [21] studied microchannels with a hydraulic diameter of 0.862–1.7 mm, and the working fluid was organic R-134a; the heat transfer coefficient under laminar flow was higher with a smaller hydraulic diameter. Park and Punch studied microchannels with a hydraulic diameter of 106–307 μm under laminar flow conditions, with a Reynolds number of 69–800 and DI water as the cooling medium; they found that the classical theory of fluids was suitable for the calculation and analysis of heat dissipation performance in this size range. Lee et al. [22] studied microchannels with a hydraulic diameter of 194–534 μm, a Reynolds number varying from 300 to 3500, and DI water as the cooling fluid; the experiment showed that, when given at the same mass flow rate, the heat transfer coefficient of the microchannel increased as the size of the channel decreased. Quan et al., who studied microchannels with a hydraulic diameter of 127–173 μm, Reynolds number of 240, and water as cooling fluid, found that under laminar flow the heat transfer coefficient of the microchannels increased with increasing fluid flow, and decreased with a diameter increase. Chen et al. studied microchannels with a hydraulic diameter of 100–250 μm, a Reynolds number of 200–1700, and water as the cooling fluid; their experiment showed that the reduction of the size of the microchannels could improve the heat transfer performance.

Mokrani et al. [23] studied microchannels with a hydraulic diameter of 1–100 μm, Reynolds number of 100–5000, and water as cooling fluid; his experimental results showed that the classical theory can be used for capability analysis of heat dissipation in microchannels with a height between 50 and 500 μm. Hiu et al. studied microchannels with an aspect ratio ranging from 1.67 to 14.29, a Reynolds number of 50–1000, and water as the cooling fluid, and showed that the heat transfer performance was more obvious in the microchannels with a large aspect ratio. Yilmza [24] studied a parallel microchannel heat sink with different cross sections and proposed an empirical formula for the relationship between the parallel microchannel structure size and heat transfer capacity. Bjorn [25] analyzed six different sizes of rectangular cross-section microchannel structures and derived the relative heat transfer formula and fluid resistance formula. Seng Lee and Garimella et al., who studied microchannels with a depth of 400 μm and a width of 102–997 μm, with DI water as the cooling medium, found that the saturation heat transfer coefficient was not sensitive to changes in heat flux. Shen et al. studied microchannels with a width of 300 μm and a height of 800 μm, with a Reynolds number of 162–1257, and DI water as the cooling medium, and found that under laminar flow conditions, the Nusselt number was significantly lower than the traditional theoretical prediction value due to the influence of the ratio of height to width and the roughness of the channel surface. Nishimura [26,27] studied arc-shaped and sinusoidal-shaped microchannels with convex-concave wall structures and found that at low Reynolds numbers, the fluid flow in the two structures was two-dimensional, but as the Reynolds number increased, the flow state changed. It was easier to enter the turbulent state at a lower Reynolds number for fluid flowing through the curved wall than for that going through the sinusoidal wall. The instability of the fluid can enhance the heat transfer effect. Sui et al. [28] discovered the influence of the amplitude and wavelength of the concave-convex inner wall surface on the heat transfer capacity. The difference in the amplitude and wavelength disturbs the flow state of the fluid. The convex-concave structure can cause the fluid flow in a vortex to destroy the fluid boundary layer, which significantly enhances the heat dissipation effect of the fluid. Li Yue et al. [29] of the University of Electronic Science and Technology of China studied convex-concave wall microchannels and analyzed the heat transfer characteristics of the concave-convex wall with different structures. The results showed that the misaligned convex-concave microchannels had lower wall temperature and thermal resistance.

Benefiting from recent progress in the research on microchannel cooling theory, efforts across academia and industry are being directed to the integration of microchannels into microwave printed circuit boards, LTCCs, and other packaging substrates to dissipate heat from the high-power RF devices assembled on them; some have already entered the commercialization stage. For example, IMST GmbH located in Germany has developed an LTCC package substrate with embedded microchannels for heat dissipation. Peking University [30] and The CETC 43 Institute demonstrated a sample of LTCC substrate with embedded microchannels in 2016. In terms of high-frequency electrical characteristics, LTCC has low high-frequency loss, high temperature tolerance, and high current adaptability, all of which make it an excellent substrate material for a multichip module in the RF field. However, the thermal conductivity of LTCC is only 2–5 W/(m·K), while Si is 150 W/(m·K) and SiC is 490 W/(m·K). Therefore there is still a large thermal resistance from the heat source of the device layer of the GaN device to microfluidics, which makes it difficult for the heat dissipation efficiency to improve. Moreover, it is difficult to further reduce the size, aspect ratio, and minimum width and spacing of RDL to fully exert the advantages of the heat dissipation technology of the LTCC embedded microchannels.

In order to explore the potential of the heat dissipation efficiency of microchannels, the US Naval Laboratory [31] comparatively studied the heat dissipation characteristics of parallel groove microchannels on Si, AlN, SiC, copper, and other substrates in 2006. The GaN device was assembled on the surface of the microchannel heat sink (the side opposite to the open channel) by using a conductive glue or eutectic bonding process, and the open surface of the microchannel heat sink was closed with a clamp. The width of the microchannel was (500 ± 25) μm, with a height of (1.4 ± 0.025) mm and a spacing of 1 mm. The GaN layer was used to make a simulated heat source and a heater was directly mounted on the surface of the heat sink to test and analyze the heat dissipation efficiency of the microchannels. The results showed that the Si and AlN microchannel cooler could make the device work in the power density range of 1000–1200 W/cm^2 (the size of the heat source was 3×5 mm^2 to 2×5 mm^2, the maximum allowable temperature of the heat source surface was 150 °C); the heat dissipation performance of the crystalline chemical vapor deposition SiC microchannel cooler was equivalent to that of the copper microchannel cooler, which could make the device work in the power density range of 3000–4000 W/cm^2 (the size of the heat source was 1.2×5 mm^2, the maximum allowable temperature of the heat source surface was 120 °C), but there was a problem of thermal expansion stress mismatch in the assembly of the copper microchannel heat sink; a layer of diamond was deposited on the basis of chemical vapor deposition of SiC, which can achieve a 20%–30% improvement in heat dissipation performance.

Compared with the SiC-based microchannel heat sink, the Si material has a lower thermal conductivity. Under the same conditions, the SiC-based microchannel has a better heat dissipation performance, but the manufacturing technology of Si-based material is more mature, which can realize the cooling function and 2.5D/3D heterogeneous integration enabled by TSV technology at the same time. In 2014, the author team

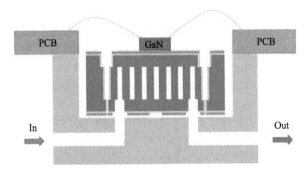

Fig. 6.2 Conceptual diagram of 2.5D integrated GaN device cooling by embedded microchannel.

[32,33] began to engage in research on TSV interposer embedded microchannels for cooling of 3D heterogeneous RF integration applications, which has greatly reduced the thermal resistance from the heat source of the GaN device to the cooling microfluid and demonstrated an embedded microchannel and redistribution layer having smaller feature size with higher precision than that of the LTCC embedded microchannel technology (Fig. 6.2).

To further reduce the thermal resistance between the heat source and the cooling microfluidics, research efforts have been directed toward extending the cooling microfluidics inside the substrate of the high power GaN chip to make a direct cooling on the heat source-active device layer, which has become a cutting-edge research topic currently. In 2015, Raytheon research team [4,34–36] proposed a GaN device with embedded microchannels made inside a solder-bonded diamond substrate and silicon substrates, aiming to further reduce the thermal resistance between the heat source-active device layer of the GaN and the cooling microfluid, and improve the heat dissipation efficiency. However, this solution has to address issues regarding the GaN-on-diamond process first; whether the GaN device layer is grown directly on the diamond substrate or the device layer of GaN is transferred to a diamond substrate, the technology readiness should be further improved. Secondly, the fabrication of a microchannel on a diamond substrate is still challenging, and the bonding process to a silicon substrate with microchannels, whether direct bonding or solder bonding, will induce thermal stress issues.

In 2016, Lockheed Martin [10,37–39] proposed a method of mounting a SiC-based GaN device that is locally thinned in a SiC substrate beneath the active area onto a manifold heat sink, where the locally thinned in SiC substrate is about 30 μm, while the whole thickness of the SiC substrate is about 100 μm. When it works, the cooling medium is sent up through the manifold heat sink and sprayed directly to the locally thinned SiC substrate. Their study shows that the gain of the GaN monolithic microwave integrated circuit (MMIC) amplifier can be increased by more than 4 dB, and its maximum output power can be increased by more than 8 dB, an improvement of about 3%–5%.

In summary, among the three solutions, the external fluid or microfluidics cooling has a high degree of maturity and is currently being applied in advanced electronic information equipment. However, the thermal resistance between the heat source of the high-power device and the heat sink is relatively large, and the heat dissipation capacity is the worst. The solution of GaN device integrated microchannel cooling has shown the great potential of improvement in heat dissipation capacity, cooperative research has to be conducted among GaN device design, process, integration, or even module assembly, as the introduction of cooling microfluidics into the chip substrate inevitably poses risk to the GaN device. Moreover, it is more a solution for heat dissipation of a monolithic GaN device; it is difficult to extend it to a heterogeneous integration scenario. Regarding the cooling solution by microfluidics inside a package, the capacity for heat dissipation is in the middle. The technology of LTCC embedded microchannels has been entering the field of advanced electronic information equipment. Compared with the technology of LTCC embedded microchannels, TSV interposer embedded microchannels have high compatibility with current electronic assembly technology, show great potential for improvement in terms of microchannel feature size, electrical wiring capability, and heat dissipation capability, and have good prospects for application. Therefore this chapter focuses on introducing this solution in conjunction with the author's scientific research practice.

6.2 Design of a HR-Si interposer embedded microchannel

The design of the 3D heterogeneous RF integration oriented HR-Si interposer embedded microchannel includes the design of the main structure of the microchannel area, the diversion area, and the TSV interconnection. Among them, the structural design of the main channel area primarily starts with a study on the characteristics of the heat source in combination with the derived required heat dissipation capacity, and then to optimization of microchannel structure in combination with the cooling medium pump pressure. The main purpose of the structural design of the diversion area is to ensure the expected fluid flow in the main microchannel region, while processing requirements should be taken into account at the same time. The TSV interconnection design mainly includes electrical interconnections through the substrate to provide electrical grounding and electrical signal transmission.

Research on the theory of microchannel cooling shows that the heat dissipation capacity can be improved by reducing the size of the microchannel, increasing the aspect ratio of the microchannels, and enhancing the turbulence mechanism. Considering the process compatibility issue at the same time, the author's team mostly studied parallel microchannel structures and periodic staggered structures, as shown in Fig. 6.3, including parallel structures, S-shaped microchannels, staggered microchannels with rectangular columns, and staggered microchannels with cylindrical columns. The width/spacing

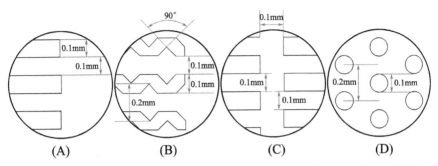

Fig. 6.3 Four kinds of main channel region design: (A) Parallel straight microchannel; (B) S-shaped microchannels; (C) staggered microchannel with rectangular columns; and (D) staggered microchannel with cylindrical columns.

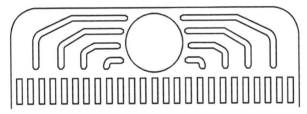

Fig. 6.4 Diversion structure design at the inlet and outlet region of the microchannel.

of the microchannel is 100 μm/100 μm, the depth is 300 μm, the aspect ratio is 3, the size of the HR-Si interposer is 9.45 mm × 9.5 mm, and the thickness is 550 μm.

Fig. 6.4 shows a schematic diagram of an optimized diversion structure, which is symmetrically arranged at the inlet and outlet region of the main microchannel to provide the required flow inside the main flow channel. Fig. 6.5 shows the simulation results of the velocity and pressure distribution from a staggered microchannel with rectangular columns with or without the design of diversion structure under the condition that the velocity of the coolant is set at 1.7 m/s. It can be found that the velocity distribution is more uniform in the main microchannel area with a diversion structure.

The 3D heterogeneous RF integration oriented HR-Si interposer embedded microchannel has a large thickness. Considering the process compatibility of the microchannel structure and the TSV structure, a stepped through-hole design composed of coaxial TSVs is proposed as shown in Fig. 6.6; from bottom to top, the diameters are set to 100, 180, and 220 μm, respectively, and the total height is 500 μm. The depth of different diameters can be adjusted according to the actual situation. The through-hole interconnection is realized by performing DRIE etching and a silicon-silicon bonding process, where the order of DRIE etching needs to be considered to ensure accurate etching depth and surface cleanliness. This design avoids the issues of directly etching high aspect ratio TSV on silicon-silicon bonded substrates, such as the influence of bonding surface oxide layer, control of the profile, and the microscopic morphology inside a via.

160 TSV 3D RF Integration

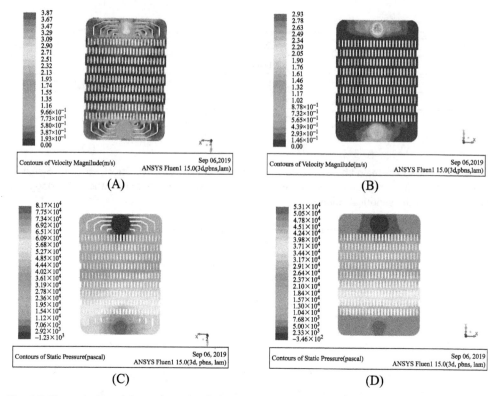

Fig. 6.5 The velocity and pressure cloud diagram with or without diversion structure under the condition of a velocity of 1.7 m/s.

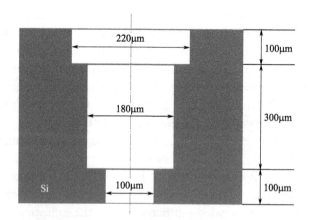

Fig. 6.6 Schematic of a stepped through-hole design composed of coaxial TSVs.

6.3 Thermal characteristics analysis of a TSV interposer embedded microchannel

In electronic packaging, the chip junction temperature or thermal resistance is used to measure the thermal performance of the package. In the application of a 3D heterogeneous RF integrated TSV interposer embedded microchannel, the high power chip (heat source) is installed on the TSV interposer, and then mounted on a carrier. The heat generated by the chip is transferred by heat conduction to the inner wall of the embedded microchannel, and then is taken away by the cooling medium. In general, the size of the RF power chip is smaller than the size of the TSV interposer, as is shown in Fig. 6.7. That means the heat is transferred from the small-area chip to a large plate, a typical heat diffusion process, resulting in a diffusion thermal resistance. When the heat source is much smaller than the size of the substrate, the diffusion thermal resistance is much larger than the equivalent thermal resistance calculated by one-dimensional conduction. Therefore the calculation and analysis of the diffusion thermal resistance are essential.

As early as the 1960s, Kennedy [40] studied the thermal diffusion effect of a small heat source on a cylindrical substrate, established a method for solving the thermal diffusion characteristics of a small heat source in the case of a cylindrical substrate, and analyzed the influence of the size of the heat source and the substrate. Based on this research, Kadambi [41] considered the convective heat transfer boundary conditions, obtained a three-dimensional steady-state thermal diffusion model with a convective heat transfer boundary at the bottom, and established a three-dimensional diffusion thermal resistance expression with a convective heat transfer boundary. John et al. [42] studied the three-dimensional steady-state thermal diffusion model of a small heat source on a rectangular substrate and obtained the thermal diffusion characteristics of the rectangular structure related to the temperature of the heat source. Yovanovic et al. [43] analyzed the thermal diffusion characteristics of a rectangular heat source on a rectangular substrate composed of two layers of different materials and obtained the corresponding thermal diffusion model. Influence of substrate thicknesses, different substrate sizes, and heat source location changes were analyzed to get the expression of diffusion thermal resistance with the corresponding parameters. Muzychka et al. [44–47] studied the thermal diffusion characteristics of multiple rectangular heat sources atop a rectangular substrate with convective heat transfer boundaries, and deduced the expressions of diffusion thermal resistance

Fig. 6.7 Schematic diagram of TSV interposer embedded microchannel integrated heat source chip.

between heat sources of different sizes and substrates. The accuracy of the expression of the diffusion thermal resistance was verified by simulation analysis, and the expression of the diffusion thermal resistance of the multilayer substrate structure was analyzed. Minseok et al. [48] studied the heat diffusion conduction process in a high-power LED array module including a fin heat sink, proposed a thermal resistance network including a fin structure, and studied the thermal diffusion under different material substrates, which was verified and analyzed by the simulation software, and the main parameters affecting the thermal diffusion process of the LED were obtained, including material properties, chip spacing, and substrate thickness. Based on Muzychka's research, Luo Xiaobing et al. [49,50] modified the heat diffusion network model under convective heat transfer conditions and verified the applicability of the model through simulation analysis. In 2016, Pi Yudan at Peking University [51] studied the heat conduction process in a TSV 3D IC and believed that the heat was mainly transferred along the high heat conduction path, such as the TSV and RDL, divided the high heat conduction path into several sections, established an equivalent thermal resistance network based on the divided sections, and the total thermal resistance could be obtained by calculating each section and adding them together. In summary, for the heat transfer problem of a small heat source on a single material or stacked substrates of different materials, a variety of calculation and analysis methods of diffusion thermal resistance have been established, including direct calculation based on analytical formulas, thermal resistance calculation based on diffusion angle, and a fitting formula based on simulation results, and equivalent thermal resistance network based on the high thermal conductivity path. Details are as follows.

6.3.1 Simplified calculation based on a variable diffusion angle

The article [52,53] proposes a simplified calculation method for thermal resistance based on a variable diffusion angle. In this method, the heat flow diffuses into the bottom plate at a constant angle, the diffusion is determined by the size of the heat source and the diffusion angle, and the size of the diffusion angle is determined by the thermal conductivity between different media. Taking a circular heat source as an example, the heat is assumed to be transferred only in the tapered region having the diffusion angle α under the temperature gradient, so that a simplified calculation formula for the thermal resistance of the multilayer dielectric structure can be obtained according to the one-dimensional equivalent thermal resistance calculation. For a heat source with a diameter of $2a$, to calculate the thermal diffusion resistance, divide the truncated cone into an infinite number of discs with a thickness of dL by the differential method. Then the equivalent thermal resistance of each disc is dL/kA, and integrate to obtain the following Eq. (6.1):

$$R = \frac{1}{k}\int_0^w \frac{dl}{A} = \frac{1}{k\pi}\int_0^w \frac{1}{(a+x\tan\alpha)^2}dx = \frac{1}{k\pi a} \times \frac{w}{a+w\tan\alpha} \quad (6.1)$$

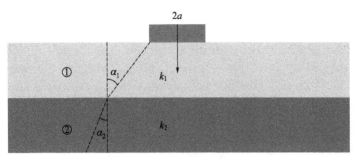

Fig. 6.8 Schematic diagram based on a variable diffusion angle.

When reaching the dielectric layer 2, the diameter of the heat source becomes the diameter when the heat source diffuses from the top to the bottom of the dielectric layer 1; that is, at layer 2, the diameter of the heat source is $2a + 2w\tan\alpha_1$. Recursively calculate the thermal resistance of each layer step by step to obtain the total thermal resistance (Fig. 6.8).

6.3.2 Direct calculation based on analytical formula

Fig. 6.9 shows a typical thermal diffusion problem, where the uniform heat source length is $2a$, the width is $2b$, the substrate length and width are $2c$ and $2d$, the thickness is t, the substrate thermal conductivity is k_s, and the bottom convective heat transfer coefficient is

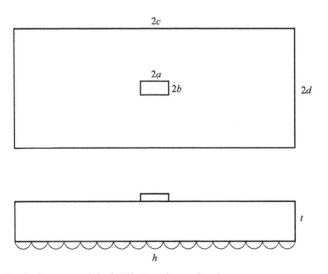

Fig. 6.9 Simplified calculation model of diffusion thermal resistance.

h, assuming that the contact thermal resistance between each layer can be ignored, the heat is only transferred out through the bottom surface flow heat transfer, and the periphery and upper surface are under adiabatic conditions. The formula of diffusion thermal resistance is derived by Refs. [44–47], based on the analytical formula is given and described in Eq. (6.2)–(6.7), the thermal resistance can be calculated directly.

$$R_{sp} = \frac{1}{2a^2 cdk_s} \sum_{m=1}^{\infty} \frac{\sin^2(a\delta_m)}{\delta_m^3} \varphi(\delta_m) + \frac{1}{2b^2 cdk_s} \sum_{n=1}^{\infty} \frac{\sin^2(b\lambda_n)}{\lambda_n^3} \varphi(\lambda_n)$$
$$+ \frac{1}{a^2 b^2 cdk_s} \sum_{m=1}^{\infty} \sum_{n=1}^{\infty} \frac{\sin^2(a\delta_m)\sin^2(b\lambda_n)}{\delta_m^2 \lambda_n^2 \beta_{m,n}} \varphi(\beta_{m,n}) \qquad (6.2)$$

$$\varphi(x) = \frac{(e^{2xt}+1)x - \rho(1 - e^{2xt})}{(e^{2xt}-1)x + \rho(e^{2xt}+1)} \qquad (6.3)$$

$$\rho = \frac{h_{equ}}{k_s} \qquad (6.4)$$

$$\delta_m = \frac{m\pi}{c} \qquad (6.5)$$

$$\lambda_n = \frac{n\pi}{d} \qquad (6.6)$$

$$\beta_{m,n} = \sqrt{\delta_m^2 + \lambda_n^2} \qquad (6.7)$$

where m and n are infinite series, and the first 500 items are selected in this article.

6.3.3 A fitting formula based on simulation results

The article [54] used different convective heat transfer coefficients in the simulation and obtained the fitting equation of the diffusion heat resistance with the change of the convective heat transfer coefficient according to the simulation results. That is, $R_{sp} = A\ln(h) + B$, and it was found that when the convective heat transfer coefficient changes from 0 to 1000 W/(m²·K), the change of diffusion thermal resistance R_{sp} has a linear relationship with the convective heat transfer coefficient h.

6.3.4 Equivalent thermal resistance network based on the high thermal conductivity path

Regarding the heat conduction process of a TSV 3D IC, the article [51] considered that the heat was mainly transferred along the high heat conduction path, such as Cu TSV and RDL, and was calculated by dividing the high heat conduction path into several sections; an equivalent thermal resistance network is thus established. Fig. 6.10 shows the typical structural unit in TSV 3D integration. Fig. 6.11 shows the corresponding differential thermal

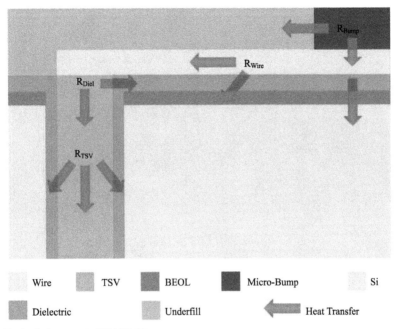

Fig. 6.10 Typical element in TSV 3D IC.

resistance network. Finally, all high heat conduction paths are calculated. The calculation methods for each component are as follows which is described in Eqs. (6.8)–(6.12):

$$R_{bump-1D} = \frac{d_{bump}}{k_{bump} A_{bump}} \tag{6.8}$$

$$R_{bump-sp} = \frac{\ln(r_{underfill}/r_{bump})}{2\pi k_{underfill} d_{bump}} \tag{6.9}$$

$$R_{wire-1D} = \frac{d_{wire}}{k_{wire} A_{wire}} \tag{6.10}$$

$$R_{TSV-1D} = \frac{d_{TSV}}{k_{TSV} A_{TSV}} \tag{6.11}$$

$$R_{TSV-sp} = \frac{\ln(r_{Si}/r_{TSV})}{2\pi k_{Si} d_{TSV}} \tag{6.12}$$

where d_{bump} means that the solder ball is divided into an infinite number of sections by differentiation, and the thickness of each section is d_{bump}; k is the thermal conductivity of the corresponding material; A is the cross-sectional area of the corresponding medium, and r is the radius of the TSV.

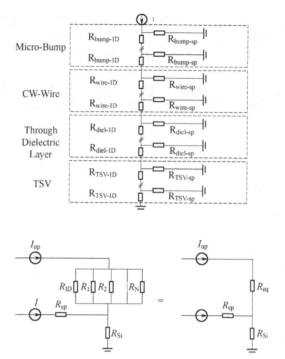

Fig. 6.11 Thermal resistance network based on high thermal conductivity path.

When a tiny heat source transfers heat to a substrate embedded with a microchannel, the structure of the microchannel is complicated, and the establishment of a thermal resistance model is more difficult. It is not only necessary to consider the influence of the structure on the heat dissipation performance, but also to consider the convective heat transfer between the fluid and the solid at different flow rates. Mao Zhangming of Huazhong University of Science and Technology [55] studied the heat transfer process of multiple heat sources on a water-cooled heat transfer structure with tree-shaped bifurcated microchannels and established a simplified thermal resistance network model of the water-cooled structure. The correctness of the model was verified by experiments and simulation analysis.

Here, the parallel microchannel is taken as an example to establish the thermal resistance model of the chip mounted on top of a high-resistivity Si interposer embedded microchannel, as shown in Fig. 6.12.

The temperature rise of the chip with reference to the coolant is expressed in Eq. (6.13):

$$T_j = T_a + R_{model} Q \tag{6.13}$$

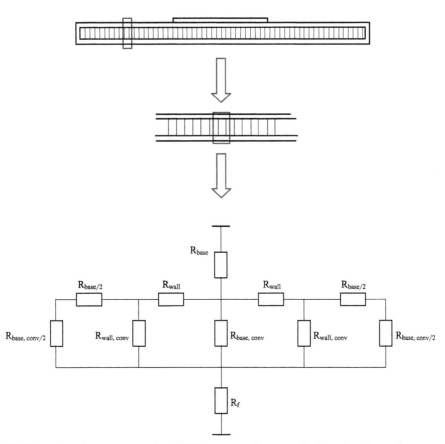

Fig. 6.12 Thermal resistance networks of TSV interposer integrated with microchannel.

where T_j is the temperature rise of the chip, and T_a is the temperature of the coolant at the inlet region. The total thermal resistance of chip1 referring to the coolant, R_{model}, is composed of the one-dimensional conductive thermal resistance of the chip itself, R_{chip}, the spread thermal resistance of chip1 to the microchannel substrate R_{sp}, the thermal resistance of the microchannel $R_{microchannel}$, and the one-dimensional thermal resistance of the substrate R_{1D-sub}, which is described in Eqs. (6.14)–(6.15):

$$R_{model} = R_{chip} + R_{sp} + R_{microchannel} + R_{1D-sub} \quad (6.14)$$

where

$$R_{chip} = \frac{t}{kab} \quad (6.15)$$

For the diffusion heat resistance R_{sp} of the chip to the substrate, Muzychka's heat sink structure is used to obtain and described in Eqs.(6.16)–(6.22) [44–47]:

$$R_{sp} = \frac{1}{2a^2 cdk_2} \sum_{m=1}^{\infty} \frac{\sin^2(a\delta_m)}{\delta_m^2} \phi(\delta_m) + \frac{1}{2b^2 cdk_2} \sum_{n=1}^{\infty} \frac{\sin^2(b\lambda_n)}{\lambda_m^2} \phi(\lambda_n) \\ + \frac{1}{a^2 b^2 cdk_2} \sum_{m=1}^{\infty} \sum_{n=1}^{\infty} \frac{\sin^2(a\delta_m)\sin^2(b\lambda_n)}{\delta_m^2 \lambda_m^2 \beta_{m,n}} \phi(\beta_{m,n}) \quad (6.16)$$

with

$$\phi(x) = \frac{(\alpha e^{4xt_1} + e^{2xt_1})x + \rho(e^{2x(t_1-t_2)} + \alpha e^{4x(t_1+t_2)})}{(\alpha e^{4xt_1} - e^{2xt_1})x + \rho(e^{2x(t_1-t_2)} - \alpha e^{4x(t_1+t_2)})} \quad (6.17)$$

$$\rho = \frac{x + h_{equ}/k_1}{x - h_{equ}/k_1} \quad (6.18)$$

$$\alpha = \frac{1 - k_1/k_2}{1 + k_1/k_2} \quad (6.19)$$

$$\delta_m = \frac{m\pi}{c} \quad (6.20)$$

$$\lambda_n = \frac{n\pi}{d} \quad (6.21)$$

$$\beta_{m,n} = \sqrt{\delta_m^2 + \lambda_n^2} \quad (6.22)$$

where m, n are infinite iterations, a and b are the length and width of the chip, c and d are the length and width of the substrate; k_1 and k_2 are the thermal conductivity of the mounted chip1 and bottom S interposer, respectively; t_1, t_2 are the thicknesses of the first layer and second layer; h_{equ} is the equivalent heat transfer coefficient and can be obtained by Eq.(6.23)

$$h_{equ} = \frac{1}{cdR_f} \quad (6.23)$$

Fig. 6.12 shows the cross-sectional view of the TSV interposer embedded with microchannels having chips mounted on its top side. To simplify the analysis, one of the microchannels is taken and its equivalent thermal resistance is established and can be calculated as Eq. (6.24) [56]:

$$R_{microchannel} = R_{base} + R_{base,conv} \prod\left(\left(R_{wall} + \left((R_{base,conv}/2 + R_{base,conv}/2)\prod R_{wall,conv}\right)\right)/2\right) + R_{fluid} \quad (6.24)$$

where R_{base} is the heat transfer resistance of the substrate, R_{wall} is the heat transfer resistance of the inside half wall of the microchannel, $R_{base,conv}$ is the heat transfer resistance due to the substrate convection, $R_{wall,conv}$ is the heat transfer resistance of the inside wall of the microchannel facing, and R_{fluid} is the thermal resistance of the coolant, which can be expressed sequentially in Eqs.(6.25)–(6.29) as follows:

$$R_{base} = \frac{t_3}{2k_{sub}(b_1 + c_1)d} \quad (6.25)$$

$$R_{wall} = \frac{a_1/2}{2k_{sub}(c_1/2)d} \quad (6.26)$$

$$R_{wall,conv} = \frac{1}{2ha_1 d} \quad (6.27)$$

$$R_{base,conv} = \frac{t_3}{2hb_1 d} \quad (6.28)$$

$$R_{fluid} = \frac{1}{\dot{m} C_p} \quad (6.29)$$

where k_{sub} is the thermal conductivity of the substrate, which is 140 W/(m·K) for Si substrate; a_1, b_1 are the width and height of the cross section of one microchannel, respectively; c_1 is the distance of adjacent microchannels; t_3 is the thickness of Si film above the microchannel; d is the length of the flow channel; C_p is the specific heat capacity of the DI water, which is 4.2×10^3 J/(kg·°C), where h is the fluid convection heat transfer coefficient, which can be calculated by Eq. (6.30), referring to the average Nusselt number [22,57]:

$$h = Nu_{ave} k_f / D_h \quad (6.30)$$

where k_f is the thermal conductivity of the fluid, which is 0.613 W/(m·K) for DI water; $\dot{m}$ is the mass flow rate in the microchannel and can be expressed as $\dot{m} = \rho v s$, where ρ is the fluid density; v is the working fluid velocity, and s is the cross-sectional area of the microchannel. D_h is the feature size. For the average Nusselt number, refer to Eqs. (6.31)–(6.36), where α is the aspect ratio.

$$Nu_{ave} = \frac{1}{C_1(x^*)^{C_2} + C_3} + C_4 \quad 1 \leqslant \alpha \leqslant 10, \ x^* < x^*_{th}, \ x^* = x/(Re \cdot Pr \cdot D_h)$$
$$1 \leqslant \alpha \leqslant 10, \ x^* < x^*_{th}, \ x^* = x/(Re \cdot Pr \cdot D_h) \quad (6.31)$$

with

$$x_{th}^* = -1.275 \times 10^{-6}\alpha^6 + 4.709 \times 10^{-5}\alpha^5 - 6.902 \times 10^{-4}\alpha^4$$
$$+ 5.014 \times 10^{-3}\alpha^3 - 1.769 \times 10^{-2}\alpha^2 + 1.845 \times 10^{-2}\alpha + 5.691 \times 10^{-2} \quad (6.32)$$

$$C_1 = -2.757 \times 10^{-3}\alpha^3 + 3.274 \times 10^{-2}\alpha^2 - 7.464 \times 10^{-5}\alpha + 4.476 \quad (6.33)$$

$$C_2 = 6.391 \times 10^{-1} \quad (6.34)$$

$$C_3 = 1.604 \times 10^{-4}\alpha^2 - 2.622 \times 10^{-3}\alpha + 2.568 \times 10^{-2} \quad (6.35)$$

$$C_4 = 7.301 - 1.311 \times 10/\alpha + 1.519 \times 10/\alpha^2 - 6.094/\alpha^3 \quad (6.36)$$

Based on the established network of thermal resistance model and parameter calculation method, Table 6.1 shows the thermal resistance of each part by calculating when the heat flux density of the power chip is $300 \, W/cm^2$ and the velocity of coolant is set to $10 \, mL/min$. Fig. 6.13 shows the maximum temperature rise of the chip surface with a changing input heat flux power at different flow rates. To verify the effectiveness of

Table 6.1 Calculated value of thermal resistance of each part.

	$R_{chip-1D}$	$R_{diffusion}$	R_{sub-1D}	R_{case}	$R_{convection}$	R_{total}
Thermal resistance (K/W)	0.2085	1.1563	0.0081	0.0175	1.8528	3.2432

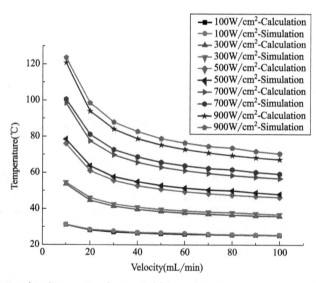

Fig. 6.13 The relationship between velocity and bottom heat source temperature under different power densities.

the calculation method, the simulation results based on Fluent finite element modeling are also shown in Fig. 6.13. It is assumed that the velocity in the microchannel is uniformly distributed in the theoretical calculation. It can be found that the calculation method based on the thermal resistance network is consistent with the simulation results. As the velocity of coolant gradually increases, the surface temperature of the power chip gradually decreases, and finally stabilizes under a given heating power input condition. For example, when the heat flux density is 100 W/cm^2, the flow rate is increased from 10 to 20 mL/min, the heat source temperature is reduced by 3.83 °C, and when the flow rate is increased from 90 to 100 mL/min, the heat source temperature is only reduced by 0.22 °C. This implies that there is a given saturation state for a given microchannel.

To further study the relationship between the effect of size of heat source on the surface temperature, Fig. 6.14 shows the surface temperature of the heat source having a changing size under input heat power. The velocity of DI coolant is set to 40 mL/min, which is approximately considered to reach the saturation state. In the case of constant input power, for example, when the heat source area is reduced by four times from 6×4.6 mm^2 to 3×2.3 mm^2, the power density will be expanded from 100 to 400 W/cm^2, and the corresponding temperature will increase from 48.96 to 68.77 °C. This means that the reduction in the area of the heat source leads to the hot spot effect, which increases the difficulty of heat diffusion to the bottom layer and the surrounding area and increases the diffusion heat resistance and the surface temperature of the heat source. This issue should be considered in thermal management design.

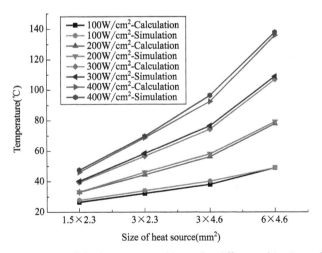

Fig. 6.14 Surface temperature of the heat source chip under different chip sizes of heat source and different power densities.

6.4 Process development of a TSV interposer embedded microchannel

Fig. 6.15 shows the process design of the HR-Si interposer embedded microchannel. At first, the microchannel structure with a depth of 300 μm is formed on the surface of a given Si using the DRIE process; similarly, a fluid inlet and outlet region with a diameter of 1 mm is formed on another Si wafer. The two wafers are aligned and directly bonded together to form the microchannel; in its body, thinning is optional, as shown in Fig. 6.15A. Secondly, the DRIE process is used to disclose the buried via inside and form a through-hole, as shown in Fig. 6.15B. Until this step, the process for the HR-Si interposer described in Section 6.2 is utilized to form an HR-Si interposer embedded microchannel.

The most challenging process for a TSV interposer embedded microchannel is the silicon-silicon bonding process. The wafer-level silicon-silicon bonding process is

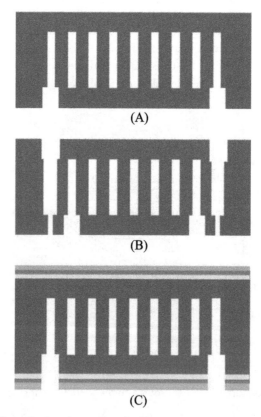

Fig. 6.15 The process flow of microchannel production.

realized with a prebonding by jointing two silicon-based wafers that have undergone standard inorganic cleaning under given temperature and pressure conditions, and a following annealing treatment that lasts several hours. A polymerization reaction occurs at the bonding interface to form a stable silicon-oxygen covalent bond to form a permanent bond, as shown in Eq. (6.37). The silicon-silicon bonding process requires extremely high surface cleanliness and flatness. However, in the current situation, microchannels on the surface will undergo bonding, which makes it hard to implement.

$$\begin{array}{c} Si- \\ Si-O \end{array} + H_2O \rightarrow \begin{array}{c} Si-OH \\ Si-OH \end{array} + H_2O \rightarrow Si-OH \cdot H_2O$$

$$Si-OH + HO-Si \rightarrow Si-O-Si + H_2O \qquad (6.37)$$

Before the Si—Si wafer bonding process, cleaning is a must step. Standard RCA cleaning is performed to remove organic matter and particles such as photoresist and dust particles. With a standard RCA cleaning, there are nonbridging hydroxyl groups on the surface of the silicon wafer, which is key for a successful bonding. The Si—Si wafer bonding process is conducted in our lab with AML bonder, the bonding chamber is set to 1×10^4 mBar, the bonding temperature is set to 100°C, the bonding pressure is 1500 N, Fig. 6.16 shows a jointed Si—Si wafer after the prebonding process. The annealing is conducted with nitrogen protection; the annealing temperature curve is shown in Fig. 6.17. Fig. 6.18 shows a jointed Si—Si wafer finishing annealing. To evaluate the quality of Si—Si bonding, ultrasonic imaging technology is used. Fig. 6.19 shows the measured ultrasonic scanning images of the bonded Si—Si wafer with microchannels. Fig. 6.20 shows a Si interposer, which is metallized with the Cu/Ni/Pd/Au layer. The SEM results of the cross section of the interposer after bonding are shown in Fig. 6.21. Fig. 6.22 shows the cross section of the coaxial TSV stepped via structure.

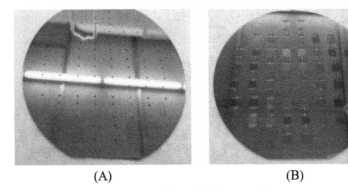

Fig. 6.16 Prebonded sample: (A) upper wafer and (B) lower wafer.

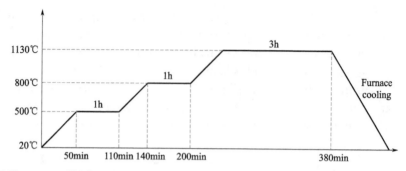

Fig. 6.17 The curve of high-temperature annealing.

Fig. 6.18 The sample of silicon-silicon bonding.

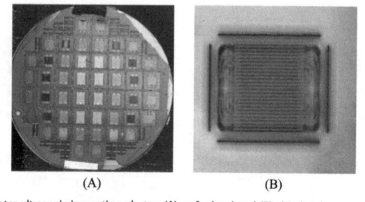

Fig. 6.19 Water ultrasonic inspection photos: (A) wafer level and (B) chip level.

HR-Si interposer embedded microchannel 175

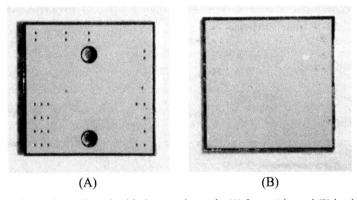

Fig. 6.20 Physical samples with embedded microchannels: (A) front side and (B) back side.

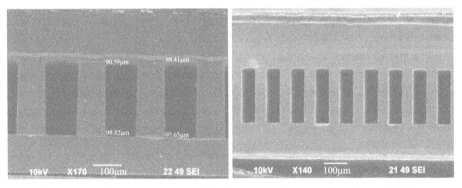

Fig. 6.21 The picture of the cross section of the interposer after bonding through SEM.

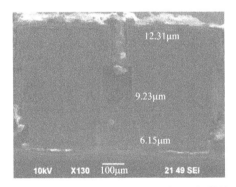

Fig. 6.22 The picture of TSV interconnection cross section through SEM.

6.5 Characterization of cooling capacity of HR-Si interposer with an embedded microchannel

Characterization of the heat dissipation capacity of the microchannel inside an HR-Si interposer is indispensable for the development of the microchannel cooled 3D heterogeneous RF integration [58]. According to the published literature [59], it is mainly derived from the measurement of heat source power, temperature, flow rate, and pressure drop of coolant. Fig. 6.23 shows a typical diagram for the measurement, including heat source, microchannel, test fixture, circuit board, pump, pressure measurement system, and temperature acquisition system.

There are several forms of heat source: an external cartridge as heat source such as the Naval Research Laboratory used in Ref. [31], dummy heat source chips, or GaN high-power devices. The heat source chip, microchannel, and test fixture are assembled with conductive glue, solder, nanosilver paste bonding or eutectic welding; a variety of circuit board is used to fan-out the signal of the heat source chip signal, where wire-bonding is used to realize the electrical connection between them.

The test fixture is generally made of aluminum alloy, stainless steel, or acrylic. There is a channel inside the body to provide fluid input and output and the necessary pressure test point. The pump can be a peristaltic pump or a gear pump; the pressure test system is mainly a pressure sensor, which is used to test the pressure difference between the input and output regions. For the temperature test, thermocouples and infrared real-time thermal imaging cameras can be used, or an on-chip temperature sensor is also a possible method. DI water, ethylene glycol, or HFE-7100 are popular coolants. The pump drives

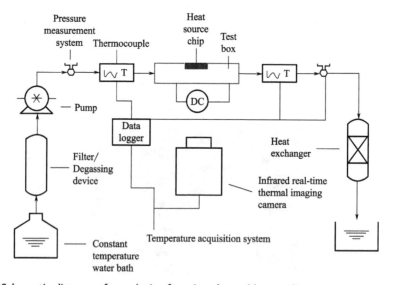

Fig. 6.23 Schematic diagram of test device for microchannel heat sink.

coolant into the microchannel through the test fixture to take away heat. The temperature of the circulating coolant will increase during the test, and a constant temperature water bath or heat exchanger can be used. A DC power source provides the supply voltage [60].

The heat dissipation capacity is generally characterized in terms of the maximum dissipation power and the maximum heat dissipation density when the chip is in normal work mode, which implies the junction temperature should remain at an acceptable level in the measurement. During the measurement, the flow rate (V), pressure (P), temperature (T), input voltage (V_n), input current (I_n), and other values are recorded. For a real chip, modeling and simulation may be needed to predict whether the junction temperature is at a safe level from the tested surface temperature. At this precondition, the maximum heat dissipation power is characterized by $P_n = V_n I_n$. The maximum heat dissipation density can be extracted by divided by the area of the heat source. Here, the input power is overestimated a little, because of the heat exchange between the chip and environment. In addition, the dissipation during the path from source to coolant should also be factored in. A popular method is to extract the dissipation without coolant at a given input power.

Also, the maximum dissipation power can be characterized in Eq.(6.38) by calculating the power taken by the coolant.

$$Q_n = \frac{(T_o - T_i)CV\rho}{A} \tag{6.38}$$

where $T_o - T_i$ represents the temperature difference between the inlet and outlet, C is the specific heat capacity of the cooling liquid, and ρ is the density of the cooling liquid. A is heat source area. This method slightly underestimates the capacity, as the heat dissipation in the HR-Si interposer is excluded.

Considering the power consumed to make a given coolant flow rate (f), to better characterize the cooling efficiency of the microchannel, Δp is pressure drop, ΔT_{max} is temperature variation, the COP (coefficient of performance) is defined as an indicator, which can be described in Eqs.(6.39)–(6.41) [61]:

$$\text{COP} = \frac{Q_{max}}{P} \tag{6.39}$$

$$P = f\Delta p \tag{6.40}$$

$$\Delta T_{max} = Q_{max} R_{tot} \tag{6.41}$$

where Q_{max} is the power dissipation of the system, and P is the pump power consumed to guarantee the junction temperature of the chip below the given maximum junction temperature. The pump power can be characterized by the flow rate and the pressure drop. R_{tot} is the total thermal resistance between the junction temperature of the device and the coolant inlet temperature. To increase COP, the pressure drop, flow rate, and thermal resistance must be reduced, but these values are interdependent and need to be considered comprehensively.

6.6 Evaluation of HR-Si interposer embedded with a cooling microchannel

In order to evaluate the HR-Si interposer embedded with a cooling microchannel in terms of cooling efficiency, a custom designed chip was utilized as a heat source, which is made of a Pt wire resistor on a $5\,mm \times 5\,mm \times 100\,\mu m$ with a metallized backside and has an on-chip temperature sensor based on the thermal sensitivity of Pt wire. Fig. 6.24 shows the assembly flow process. First, the sample of the HR-Si interposers embedded with a microchannel was assembled onto a customized cartridge with millimeter-scale channels inside, using H20E conductive silver paste. The leakage test was carried out under a maximum flow rate of $150\,mL/min$. Second, the custom designed heat source

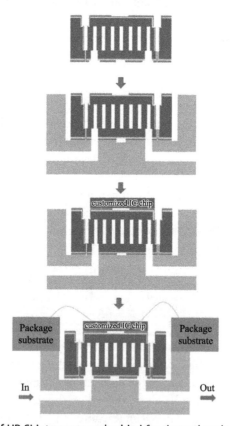

Fig. 6.24 Assembly flow of HR-Si interposer embedded for thermal evaluation.

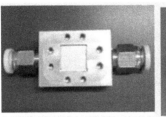

Fig. 6.25 The assembled sample for thermal evaluation.

chip was assembled to the surface of the HR-Si interposer by nanosilver cantering, which has extremely high thermal conductivity, good electrical conductivity, excellent low-temperature sintering performance, among other characteristics, and all of that can effectively be employed to reduce the thermal resistance effect between the chip and the interposer. Third, the PCB was fixed, and wire-bonding was done to achieve electrical interconnection. Fig. 6.25 shows the assembled sample for thermal evaluation.

According to the evaluation method stated in Section 6.5, DI water was used and set at a flow rate of 100 mL/min and 150 mL/min, respectively; the initial voltage was set at 20 V, incremented in steps of 5 V; resistance of the customized heat source chip, the pressure drop, and temperature difference between the inlet and outlet regions of the microchannel were recorded. The corresponding temperature of the chip surface can be obtained from the calculated value of resistance from Eq. (6.42):

$$T_{or} = \frac{R_T - R_0}{T - T_{ref}} \times \frac{1}{R_0} \times 10^6 \tag{6.42}$$

where T_{or} is the temperature coefficient of the customized heat source chip; this value is extracted experimentally; T_{ref} is the reference temperature, which is room temperature or 25 °C; T is the stable chip surface temperature; R_0 is the initial resistance value of the customized heat source chip, which is 1044.2 Ω; R_T is the resistance value recorded during the test.

Fig. 6.26 shows the deviation between the test results from the FLIR imaging technology and the theoretical result from the thermal sensitivity of a Pt wire resistor. The former is extracted from the whole surface temperature of the heat source. During the test, the entire test surface can be sprayed with black paint to calibrate the emissivity for optimization. The latter is in direct contact with the heat source, and the precision can be improved with compensation. In this experiment, the maximum temperature difference is 3.5 °C, and the minimum temperature difference between them is 0.44 °C.

Fig. 6.27 shows the relationship between the maximum temperature and input heat flux when the DI coolant is set at a flow rate of 100 mL/min in the experiment, which shows a good linear relationship, but the value of the measurement is much higher than that derived from simulation. Fig. 6.28 shows the IR image when the heat flux density is

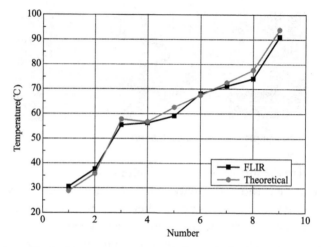

Fig. 6.26 Comparison of actual test temperature and theoretically derived temperature.

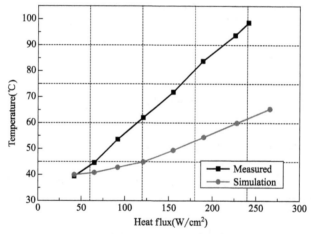

Fig. 6.27 Comparison of measured and simulated heat flux and temperature at a flow rate of 100 mL/min.

raised to 283 W/cm^2; the DI water coolant is set at a flow rate of 150 mL/min, where the temperature in the lower right corner is abnormally high. If we remove the highest temperature in the lower right corner as an error value and re-record the average value of the highest surface temperature with the software, the results are as summarized in Fig. 6.29; the gap between simulated and measured is within 10 °C. The reason for the deviation may be bubbles in the interface between the heat source chip and HR-Si interposer; further investigation also supports this point. Therefore sufficient care should be paid to the assembly of the heat source, and the thermal resistance contributed by the interface

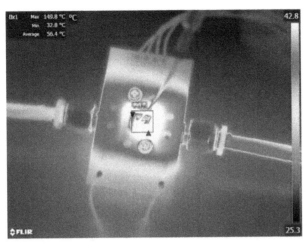

Fig. 6.28 A captured IR image of heat source chip when the heat flux density is raised to 283 W/cm² with DI water at a flow rate of 150 mL/min.

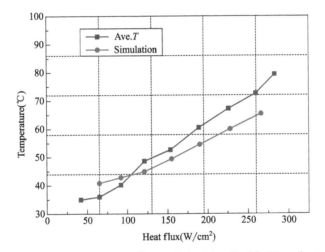

Fig. 6.29 Relationship between input heat flux and the Max T with DI coolant at a flow rate of 100 mL/min based on the measured and simulated results.

should not be ignored, especially in research and development of cooling of a heat source with very high heat flux.

To exclude thermal resistance of a uni-deal interface between heat source chip and Si interposer due to the jointing process of the conductive silver and derive the characteristics of the microchannel, a simulated heat source made of a Pt wire resistor is formed on the surface of the Si interposer directly. Fig. 6.30 shows the fabricated heat source. In a similar way, the Si interposer with an embedded microchannel is assembled. Fig. 6.31 shows the assembled sample.

Fig. 6.30 Micrograph of Pt wire resistor.

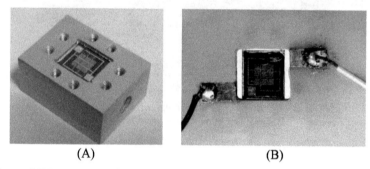

(A) (B)

Fig. 6.31 Photo of Si interposer with an embedded microchannel finishing assembly.

Fig. 6.32 shows the experimental setup. A gear pump is used to pump DI water as the cooling medium through the Si interposer embedded microchannel to take away heat. The pressure sensors are used to test the pressure changes between the inlet and outlet region under different flow rates; the heat generated by the chip is controlled by adjusting the input voltage of the thermal source chip. A K-type thermocouple and an infrared

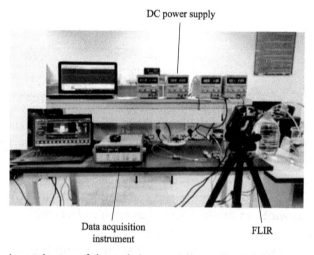

Fig. 6.32 The experimental setup of thermal characteristics evaluation.

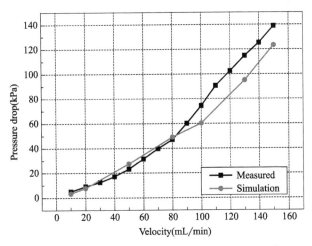

Fig. 6.33 The relationship between the pressure drop and the velocity of DI coolant with test and simulated result.

camera are used to test the surface temperature of the dummy chip at the same time for calibration.

Fig. 6.33 shows the relationship between the pressure drop and the velocity of DI coolant with test and simulated results. Table 6.2 compares the test temperature with the thermal imaging camera and thermocouple sensor, respectively, when the input power is set to 7.6 W. As the velocity of the DI coolant increases from 10 to 80 mL/min, the temperature captured is reduced to a stable value gradually, which is in agreement with the theoretical prediction in Section 6.2. This implies that for the given microchannel design, the coolant flow rate should be raised to guarantee the capacity of cooling microfluidic entering a saturation state in a practical application. In this state, it offers its best capacity and a stable cooling capacity in a relatively large range of input power.

Table 6.2 Comparison of FLIR and thermocouple test for temperature.

Velocity (mL/min)	Temperature of FLIR (°C)	Temperature of thermocouple (°C)	Deviation
10	32.3	33.7	1.4
20	28.9	30.8	1.9
30	27.7	29.4	1.7
40	27	28.8	1.8
50	26.7	28.3	1.6
60	26.3	27.9	1.6
70	26	27.6	1.6
80	25.7	27.5	1.8
Average			1.675

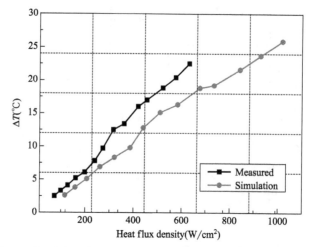

Fig. 6.34 Comparison of simulated and measured for relationship of thermal flux density and temperature difference between inlet and outlet at a flow rate of 80 mL/min.

A stable difference of less than 2 °C is found between them. This is mainly because the temperature acquired by the IR image is the extracted maximum temperature value from the whole temperature contour, while that from the thermocouple sensor is only one point of a heat source. This deviation can be compensated for. If we deduct the stable deviation between them, the error is about 0.5, which proves that both methods are acceptable. Fig. 6.34 shows the relationship between the input heat flux density and the temperature difference between the inlet and outlet region when the coolant is set at a flow rate of 80 mL/min. Fig. 6.35 shows the relationship between the input heat flux density and the maximum surface temperature. Fig. 6.36 shows the captured IR image when the heat flux density is raised to 865.24 W/cm² with a DI coolant flow rate of 80 mL/min; the maximum surface temperature is 74.1 °C. Compared with the two test curves, we can find that the temperature difference has a similar trend with the surface temperature as the heat flux density increases, while the temperature difference between the inlet and outline region is a little fewer than the variation of the maximum surface temperature, the maximum deviation is about 10°C; this difference accounts for the thermal loss due to heat exchange with air and heat transfer in the path. This is in agreement with the analysis in Chapter 6.3. Compared with the simulated results, a changing gap between them is found, as shown in Figs. 6.34 and 6.35. When the input heat flux density is less than 400 W/cm², the measurement temperature is basically lower than the simulation results, while when the heat flux density is higher than 400 W/cm², the measurement is higher than the simulation results. This is mainly because the simulation does not factor in the aluminum alloy box. In the experiment, when the input heat flow density is small, the aluminum alloy box acts as a thermal sink and benefits heat dissipation; when

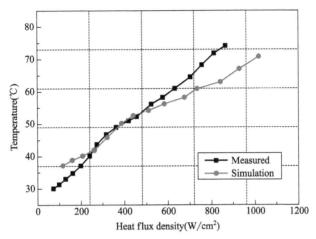

Fig. 6.35 Comparison of simulated and measured for relationship of thermal flux density and the maximum temperature at a flow rate of 80 mL/min.

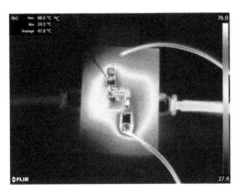

Fig. 6.36 Captured FLIR image of dummy chip with the DI flow rate of 80 mL/min under an equivalent thermal flux density of 865.24 W/cm^2.

the input heat flow density is further increased, the heat dissipation capacity of the aluminum alloy box will reach saturation, hindering the heat dissipation. This point can be proven by the IR image in Fig. 6.36.

Based on the analysis given here, the four kinds of microchannels designed are compared in terms of cooling capacity and pumping power when the DI water is set at 80 mL/min to guarantee entering a saturation state, summarized in Figs. 6.37 and 6.38. Among them, the staggered microfluidics with cylindrical column microchannel has the best heat dissipation capacity, followed by the staggered microchannel with rectangular columns, and the parallel microfluidics type is the worst. Regarding the power consumed at a given flow rate, the parallel performs best, followed by the staggered microfluidics with rectangular column microchannel, then the staggered microfluidics with cylindrical

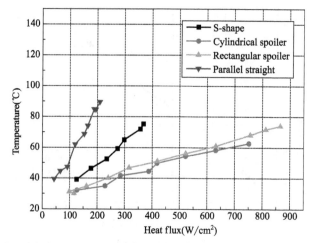

Fig. 6.37 Comparison of heat dissipation capacity of different microchannel configurations.

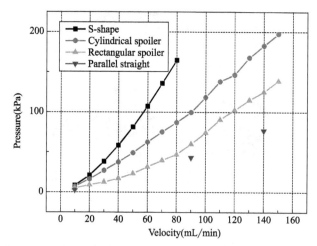

Fig. 6.38 The relationship between velocity and pressure drop of different microchannel configurations.

column microchannel, and finally the S-shaped type. From the view of manufacturability, the parallel is easier than the other three kinds.

Referring to the method described in Section 6.5, the COPs at different working conditions are calculated and compared for the four proposed microfluidic configurations. Table 6.3 shows the COPs of microchannels in our experiment with DI water having a set flow rate of 80 mL/min. One thing needs to be clarified: although the temperature rise is still at an acceptable level, the input power is not raised further to explore its ultimate cooling capacity here. This is because the electrical connection between the heat source

Table 6.3 The COPs of microchannels in our experiment with DI water having a set flow rate of 80 mL/min.

Structure	Temperature rise (K)	Velocity (mL/s)	Pressure (kPa)	Heat flux (W/cm^2)	COP
Parallel	39.5	1.33	23	120	982
S-shaped	50	1.33	164.64	368	303
Cylindrical columns	37.7	1.33	50.283	752	1083
Rectangular columns	49.1	1.33	46.983	865	3261

Table 6.4 The COPs of microchannels predicted with a set temperature rise of 60 K.

Structure	Temperature rise (K)	Velocity (mL/s)	Pressure (kPa)	Heat flux (W/cm^2)	COP
Parallel	60	1.33	23	196	1503
S-shaped	60	1.33	164.64	468	386
Cylindrical columns	60	1.33	50.283	1319	2119
Rectangular columns	60	1.33	46.983	1170	3937

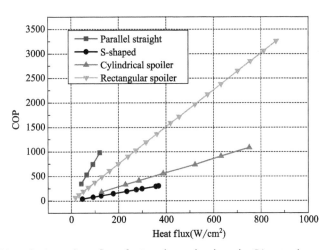

Fig. 6.39 The COP vs the input heat flux of microchannels when the DI water has a same flow rate of 80 mL/min.

chip and PCB board suffer a fuse due to big current. Table 6.4 the COPs of microchannels predicted when with a set temperature rise of 60 K. Fig. 6.39 shows the COP versus the input heat flux of microchannels when the DI water has the same flow rate of 80 mL/min. We can conclude that when the power of the chip is high, considering the COP index, the staggered microchannel with rectangular columns can maintain the balance between the heat dissipation capacity and the pump power at the same time, to obtain the best comprehensive performance. When the chip power is low, the COP index of the parallel microchannel structure is higher, and it is more economical.

6.7 Application verification of HR-Si interposer embedded with microchannel

Fig. 6.40 shows the assembly flow of a 2.5D integrated GaN power amplifier module based on an interposer embedded with a cooling microchannel. First, a customized aluminum alloy box with millimeter-level fluid channels is prepared. A molybdenum copper base is fixed, then an HR-Si interposer embedded with a cooling microchannel is soldered on the molybdenum copper base using eutectic welding. Then, the GaN power amplifier is mounted on the surface of the HR-Si interposer with gold-tin eutectic welding, and wire-bonding follows. Finally, passive devices and other necessary

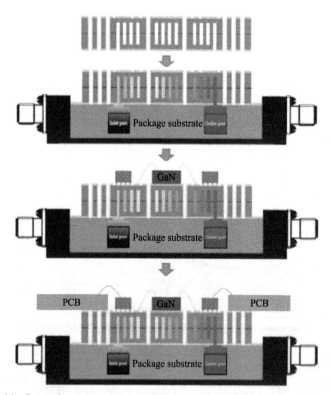

Fig. 6.40 Assembly flow of a 2.5D integrated GaN power amplifier module based on an HR-Si interposer embedded with a cooling microchannel.

Table 6.5 The tests during the sample assembly.

No.	Step	Item	Purposes
1	Before assembly	Temperature storage experiment	Check the state of the metallization of HR-Si interposer
2	After HR-Si interposer is welded to MoCu based in the customized box	Air tightness test Temperature shock test	Check the welding quality and hermeticity
3	Finished module assembly	Reliability test	Evaluate the reliability of the assembly before functional testing

microelectronics chips are mounted on, sequentially, microwave printed electrical board or other kinds of electrical boards made on ceramic, followed by a wire-bonding process or soldering to realize electrical connection with the chips. Finally, a metal cap is covered in the box to ensure sufficient chip stability.

During the assembly, a series of tests are required. Table 6.5 shows the test during the sample assembly process, including:

(1) *Temperature storage experiment.* Store the HR-Si interposer embedded with a cooling microchannel in a high-temperature environment of 300°C for about 30 min, and check whether the metal layer on the surface of the sample is bulging, peeling, etc.

(2) *Air tightness test.* After welding the high-R Si interposer to the customized metal box, an air-tightness test is conducted to ensure good welding and hermeticity.

(3) *Temperature shock test.* Carry out the cyclic temperature shock test in the temperature range of 65–150°C for the assembly parts. The purpose is to test the reliability of welding and bonding. After the temperature shock, the air tightness test is still required to ensure that the sample maintains hermeticity.

(4) *Reliability test.* After mounting of the GaN power amplifier and other passive components on the surface of the HR-Si interposer, the air tightness test and other environmental tests are conducted again to ensure the reliability of the final assembly.

Fig. 6.41 shows the assembly details of a 2.5D integrated GaN power amplifier module based on an HR-Si interposer embedded with a cooling microchannel.

Functional verification is carried out with the preceding 2.5D integrated GaN power amplifier module. DC power is applied to the module under test through a ribbon-bonded transmission line. An RF signal generator is used to control the input signal power of the module. A 30 dB attenuator is added before the amplified RF signal is sent

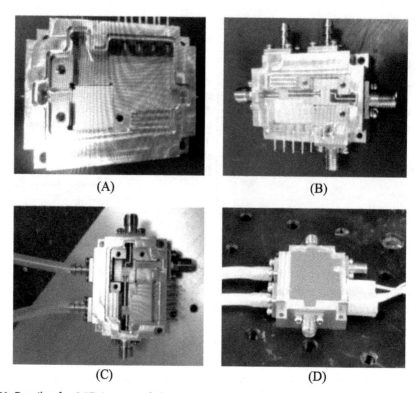

Fig. 6.41 Details of a 2.5D integrated GaN power amplifier module based on an HR-Si interposer embedded with a cooling microchannel during assembly: (A) soldering of HR-Si interposer, (B) fixing of ceramic circuit board, (C) soldering of GaN and other passive devices, and (D) capping.

to the spectrum analyzer, as the saturated output power of the GaN power amplifier used in the module here is 43 dBm in the 2–6 GHz frequency band. In this experiment, the DI water is used as the coolant, and the flow rate is set to 80 mL/min. The GaN power amplifier under test is able to increase, to work at 42–43 dB (with a 30 dB attenuation). Fig. 6.42 shows the captured small RF signal gain in the full working band, which has a maximum variation of about 6 dB. Table 6.6 illustrates the recorded input/output power and DC current as the output power of the GaN power amplifier is tuned from 27 to 43 dBm in 2–6 GHz. Based on the experiment, it can be concluded that the 2–6 GHz GaN power amplifier module is capable of working at its saturated state without performance degradation. The total power input includes the DC power supply and RF signal input; deducting the RF signal power, the total dissipated heat can be obtained. Here, the total power is about 20 W, implying generation of an equivalent heat flux of at least 400 W/cm^2, which provides a good showcase of the potential of the microchannel cooling method.

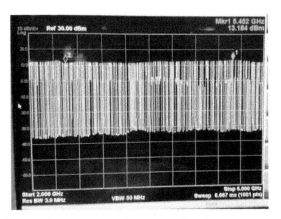

Fig. 6.42 The captured small RF signal gain in the full working band.

Table 6.6 The tested input/output power and DC current when the output power of the GaN power amplifier is increased to 43 dBm in 2–6 GHz test data.

Frequency band	Input (dBm)	Output (dBm)	Current (A)	Frequency band	Input (dBm)	Output (dBm)	Current (A)
2 GHz	−20	−3	1.73	4 GHz	−20	−3.7	1.77
	−15	1.5	1.73		−15	1.2	1.77
	−10	6.4	1.81		−10	5.7	1.85
	−5	10.5	2.07		−5	9.4	2.14
	0	11.8	2.36		2	12.6	2.85
	3	12	2.51	5 GHz	−20	−3.3	1.78
	5	12.2	2.65		−15	1.75	1.78
2.5 GHz	−20	0	1.67		−10	6.3	1.92
	−15	4.8	1.73		−5	10	2.3
	−10	8.9	1.99		0	12.6	2.88
	−5	12	2.51	6 GHz	−20	−1.7	1.73
	−3	12.5	2.69		−15	3.4	1.73
	−1	12.9	2.88		−10	7.9	1.86
3 GHz	−20	−1	1.74		−5	10.9	2.26
	−15	3.6	1.78		0	11.7	2.6
	−10	7.7	2				
	−5	11	2.56				
	−3	12.1	2.88				

6.8 Conclusions

In view of the heat dissipation problems faced by 3D heterogeneous integration of high-power GaN HEMT devices, this chapter expanded on progress made in our group on cooling solutions using HR-Si interposer embedded microchannels, including the structure design of an integrated microchannel, the thermal characteristics modeling and analysis methods for the TSV compatible process validation, evaluating methods for heat dissipation capacity. Four kinds of microchannels were proposed and compared by simulation and experimental results; the study showed that staggered microchannels with rectangular columns offered the best cooling capacity, while it suffers the biggest pressure drop, which means it consumes the most pumping power; the parallel microfluidics offers the worst cooling capacity while consuming the lowest pumping power. With the best one, a maximum heat dissipation capacity of $865.24 \, W/cm^2$ with a temperature of $74.1\,°C$ on the surface of the heat source was reached when the flow rate of DI was set to $80\,mL/min$ for the staggered microchannels with rectangular columns, which is consistent with the theoretical prediction result. On this basis, a 2.5D integrated 2–6 GHz GaN power amplifier module cooled by silicon interposer embedded microchannels was assembled and measured; the experiment shows that the 2–6 GHz GaN power amplifier works normally when its power is raised to 20 W, which means a power density of about $400\,W/cm^2$, which demonstrates the feasibility and potential for cooling of 3D heterogeneous RF integration.

The study also discusses the possibility of on-chip temperature monitoring by the characteristics of temperature sensitivity of the integrated Pt resistance wire on chip; adaptive control of coolant according to monitored temperature may be implemented.

At present, the research has not factored in the influence of the microscopic surface of the inner wall of the microchannel on the cooling performance. In the future, the cooling microfluidics will enter into the substrate of high-power devices such as GaN HEMT devices; as with the development of GaN-based high-performance TR module, cooperative research will be inevitable among GaN device design, process, integration, or even module assembly, and the reliability issues that emerge with embedded cooling microfluidics should be carefully addressed.

References

[1] Schwitter BK, Parker AE, Mahon SJ, et al. Impact of bias and device structure on gate junction temperature in AlGaN/GaN-on-Si HEMTs. IEEE Trans Electron Devices 2014;61(5):1327–34.
[2] Darwish AM, Bayba AJ, Hung HA. Thermal resistance calculation of AlGaN–GaN devices. IEEE Trans Microw Theory Tech 2004;52(11):2611–20.
[3] Chen X, Donmezer FN, Kumar S, et al. A numerical study on comparing the active and passive cooling of AlGaN/GaN HEMTs. IEEE Trans Electron Devices 2014;61(12):4056–61.
[4] Tyhach M, et al. S2-T3: next generation gallium nitride HEMTs enabled by diamond substrates. In: 2014 Lester Eastman conference on high performance devices (LEC). IEEE; 2014.
[5] Bar-Cohen A, Maurer JJ, Felbinger JG. Darpa's intra/ interchip enhanced cooling (icecool) program. In: CS MANTECH conference, New Orleans, Louisiana, USA; 2013. p. 171–4.
[6] Wei T, et al. High efficiency direct liquid jet impingement cooling of high power devices using a 3D-shaped polymer cooler. In: IEEE international electron devices meeting; 2017. p. 32.5.1–4.
[7] Glavin NR, et al. Flexible gallium nitride for high-performance, strainable radio-frequency devices. Adv Mater 2017;29:1701838.

[8] Duwei H, Miao M, Ma S, et al. Investigation of cooling performance of micro-channel structure embedded in LTCC substrate for 3D micro-system: international conference on solid-state and integrated circle technology (ICSICT), Xian, China; 2012. p. 1–3.
[9] Xia Y, Ren K, Ma S. Process development and characterization for intergrating microchannel into TSV interposer. In: Electronic components and technology conference (ECTC), Las Vegas, NV, USA; 2016. p. 2487–93.
[10] Won Y, et al. Fundamental cooling limits for high power density gallium nitride electronics. IEEE Trans Compon Packag Manuf Technol 2017;5(6):737–44.
[11] Tyhach M, Altman D, Beenstein S, et al. Next generation gallium nitride HEMTs enabled by diamond substrates. In: IEEE Lester Eastman conference; 2014. p. 1–4.
[12] Chen T, Kong Y, Wu L. Research and development of solid-state electronics. vol. 36; 2016. p. 360–4 [05] (in Chinese].
[13] Li J, Wang K. Research progress of GaN-based microwave power device based on diamond. J Xi'an Univ Posts Telecommun 2016;21(03):25–31 [in Chinese].
[14] Zhao J. Synthesis and characterization of diamond based GaN HEMT. Semicond Technol 2019;44(05):321–8. 334 (in Chinese).
[15] Jia X, Wei J, Huang Y, et al. Advances in the application of diamond cooling substrate in GaN-based power devices. Surf Technol 2020;49(11):111–23 [in Chinese].
[16] Wu B, Ye R. Research on the effect of packaging shell on high power device heat dissipation characteristics. Mach Electron 2020;2:21–4 [in Chinese].
[17] Tuckerman DB, Pease RFW. High-performance heat sinking for VLSI. IEEE Trans Electron Device Lett 1981;2(5).
[18] Bernhardt AF, Barfknecht AT, Contolini RJ, Malba V, Mayer ST, Raley NF, Tuckerman DB. Multichip packaging for very-high-speed digital systems. Appl Surf Sci 1990;46:121–30.
[19] Mundinger D, Beach RJ, Benett WJ, et al. Demonstration of high-performance silicon microchannel heat exchangers for laser diode array cooling. Appl Phys Lett 1988;53:1030–2.
[20] Peng XF, Peterso GP. Convective heat transfer and flow friction for water flow in microchannel structures. Int J Heat Mass Transf 1996;39:2599–608.
[21] Owhaib W. Experimental heat transfer, pressure drop, and flow visualization of R134a in a vertical mini/micro tube. PhD thesis, Royal Institute of Technology; 2007.
[22] Lee PS, Garimella SV. Thermally developing flow and heat transfer in rectangular microchannels of different aspect ratios. Int J Heat Mass Transf 2006;49(17–18):3060–7.
[23] Mokrani O, Bourouga B, Castelain C, et al. Fluid flow and convective heat transfer in flat microchannels. Int J Heat Mass Transf 2009;52(5–6):1337–52.
[24] Rohsenow WM, Hartnett JP, Ganic EN. Handbook of heat transfer applications. vol. 4(1). New York: McGraw-Hill Book Co; 1985. p. 266–79.
[25] Bjorn P. Heat transfer in microchannels. Microscale Thermophys Eng 2001;5(1):155–75.
[26] Nishimura T, Nakagiri H, Kunitsugu K. Flow patterns and wall shear stresses in grooved channels at intermediate Reynolds numbers. Trans JSME Ser B 1996;62(2):2106–12.
[27] Nishimura T, Kunitsugu K. Three-dimensional grooved channel flows at intermediate Reynolds numbers. Exp Fluids 2001;31(1):34–44.
[28] Y. Sui. C.J Teo, P.S Lee, et al. Fluid flow and heat transfer in wavy microchannels. Int J Heat Mass Transf, 2010, 53 (12): 2760–2772.
[29] Li Y. Study on heat dispersion characteristics of convex and convex microchannels with multiple heat sources. Chengdu: University of Electronic Science and Technology of China; 2016 [in Chinese].
[30] Hu D, Miao M, Fang R, et al. Heat transfer performance of microchannels in LTCC substrate. High Power Laser Part Beams 2016;28, 064126.
[31] Calame JP, et al. Experimental investigation of microchannel coolers for the high heat flux thermal management of GaN-on-SiC semiconductor devices. Int J Heat Mass Transf 2007;50(23–24):4767–79.
[32] Cai H, et al. Thermal and electrical characterization of TSV interposer embedded with microchannel for 2.5D integration of GaN RF devices. IEEE; 2018. p. 2156–62.
[33] Ma S, Lian T, Cai H, Hu L, He S. Thermal property evaluation of TSV interposer embedded microfluidics for cooling 2.5D integrated high power IC device. In: 2019 20th international conference on

electronic packaging technology (ICEPT), Hong Kong, China; 2019. p. 1–4. https://doi.org/10.1109/ICEPT47577.2019.245289.

[34] Altman D, Gupta A, Tyhach M. Development of a diamond microfluidics based intra-chip cooling technology for GaN. In: ASME InterPACK proceedings, USA; 2015. p. 48179.

[35] Altman D, Tyhach M, Mcclymonds J, et al. Analysis and characterization of thermal transport in GaN HEMTs on diamond substrates. In: 2014 IEEE intersociety conference on thermal and thermomechanical phenomena in electronic systems (ITherm). IEEE; 2014.

[36] Kendig D, Pavlidis G, Graham S, et al. UV thermal imaging of RF GaN devices with GaN resistor validation. In: 2018 91st ARFTG microwave measurement conference (ARFTG); 2018.

[37] Ditri J, Pearson RR, Cadotte R, et al. GaN unleashed: the benefits of microfluidic cooling. IEEE Trans Semicond Manuf 2016;29(4):376–83.

[38] Ditri J, Hahn J, Cadotte R, et al. Embedded cooling of high heat flux electronics utilizing distributed microfluidic impingement jets. In: Proc. ASME InterPACK, San Francisco, CA, USA; 2015.

[39] Ditri J, et al. Impact of microfluidic cooling on high power amplifier RF performance. In: Thermal & thermomechanical phenomena in electronic systems. IEEE; 2016.

[40] Kennedy DP. Spreading resistance in cylindrical semiconductor devices. J Appl Phys 1960;31:1490–7.

[41] Kadambi V, Abuaf N. An analysis of the thermal response of power chip packages. IEEE Trans Electron Devices 1985;32(6):289–302.

[42] John M, Krane M. Constriction resistance in rectangular bodies. Trans ASME 1991;113:392–6.

[43] Yovanovich MM, Muzychka YS, Culham JR. Spreading resistance of isoflux rectangles and strips on compound flux channels. J Thermophys Heat Transf 1999;13:428–39.

[44] Muzychka YS, Culham JR, Yovanovich MM. Thermal spreading resistances in rectangular flux channels part I—geometric equivalences. In: 36th AIAA thermophysics conference June 23–26 Orlando, Florida; 2003. p. 4176–87.

[45] Muzychka YS, Culham JR, Yovanovich MM. Thermal spreading resistances in rectangular flux channels part II—edge cooling. In: 36th AIAA thermophysics conference June 23–26 Orlando, Florida; 2003. p. 4187–95.

[46] Muzychka YS, Yovanovich MM, Culham JR. Influence of geometry and edge cooling on thermal spreading resistance. J Thermophys Heat Transf 2006;20:247–55.

[47] Muzychka YS, Culham JR, Yovanovich MM. Thermal spreading resistances of eccentric heat sources on rectangular flux channels. J Electron Packag 2003;125(2):178–85.

[48] Minseok H, Samuel G. Development of a thermal resistance model for chip-on-board packaging. Microelectron Reliab 2012;52(5):836–44.

[49] Luo XB, Mao ZM, Liu S. Analytical thermal resistances model for eccentric heat source on rectangular plate with convective cooling at upper and lower surfaces. Int J Therm Sci 2011;50(11):2198–204.

[50] Luo XB, Mao ZM, Liu S. An analytical thermal resistance model for calculating mean die temperature of typical BGA packaging. Thermochim Acta 2011;512(1):208–16.

[51] Pi Y, Wang N, Chen J, et al. Anisotropic equivalent thermal conductivity model for efficient and accurate full-chip-scale numerical simulation of 3d stacked ic. Int J Heat Mass Transf 2018;361–78.

[52] Masana FN. A new approach to the dynamic thermal modelling of semiconductor packages. Microelectron Reliab 2001;41(6):901–12.

[53] Masana FN. A closed form solution of junction to substrate thermal resistance in semiconductor chips. IEEE Trans Compon Packag Manuf Technol A 2002;19(4):539–45.

[54] Wang N. Preliminary study on thermal simulation method of 3D integrated circuit based on equivalent thermal conductivity. Shenzhen: Peking University Shenzhen Graduate School; 2017 [in Chinese].

[55] Zhang M, Luo X, Liu J, et al. J Eng Thermophys 2011;32(02):303–7 [in Chinese].

[56] Mao Z, Luo X, Liu S. Compact thermal model for microchannel substrate with high temperature uniformity subjected to multiple heat sources. In: Electronic components & technology conference. IEEE; 2011. p. 1622–72.

[57] Lei N, Skandakumaran P, Ortega A. Experiments and modeling of multilayer copper minichannel heat sinks in single-phase flow. In: Thermal & thermomechanical intersociety conference on phenomena in electronics systems; 2006. p. 9–18.

[58] Chen C, Hou F, Sub M, et al. Pressure drop and thermal characteristic prediction for staggered strip fin microchannel. IEEE Trans Compon Packag Manuf Technol 2019;99:1.
[59] Back D, Drummond KP, Sinanis MD, et al. Design, fabrication, and characterization of a compact hierarchical manifold microchannel heat sink array for two-phase cooling. IEEE Trans Compon Packag Manuf Technol 2019;99:1.
[60] Drummond KP, Back D, Sinanis MD, et al. Characterization of hierarchical manifold microchannel heat sink arrays under simultaneous background and hotspot heating conditions. Int J Heat Mass Transf 2018;126(Pt A):1289–301.
[61] van Erp R, Soleimanzadeh R, Nela L, et al. Co-designing electronics with microfluidics for more sustainable cooling. Nature 2020;585:211–5.

CHAPTER 7

Patch antenna in stacked HR-Si interposers

7.1 Introduction

When the working frequency is in the millimeter-wave band, the size of the antenna is greatly reduced, which leads to the possibility of antenna integration into the package, or even on a chip. According to the different integration methods, millimeter-wave antennas can be divided into package-integrated antennas, chip-integrated antennas, and hybrid integration [1]. Also, due to different structural features and working principles, miniaturized antennas can include patch antennas, inverted-F antennas, dipole antennas, and Yagi antennas. The types of integrated antennas have different characteristics. For example, packaged antennas and on-chip can make use of existing mature microelectronics technology. However, purely integrated antennas have higher losses in interconnection structures. On the other hand, hybrid integration technology avoids the disadvantages of packaged antennas and on-chip antennas in terms of loss, has good design flexibility, but it requires specific processes and thus higher costs.

In the field of package-integrated antennas, many well-known international research institutions such as Intel have used traditional PCB board-level integration methods based on Rogers substrate materials and advanced three-dimensional packaging methods, represented by LTCC, to carry out a series of important research works [2–7]. In 2018, Siliconware Precision Industry Co., Ltd. adopted the traditional PCB integration method, using a stacked structure and an array antenna to achieve a wider working bandwidth. The operating frequency of the antenna array is 28 GHz, and the entire stacked structure is three layers of dielectric and four layers of metal. The test result shows a relative bandwidth of 15.4%. TSMC, Micron, etc., successively reported on integrated antenna technology based on a fan-out wafer-level packaging platform. However, in the classic PCB board-level integration method, due to inherent characteristics such as accuracy and the electrical characteristics of the board, high-frequency devices have a larger size and high dielectric constant. LTCC, as a mature high-frequency interconnect substrate technology with the high dielectric constant of ceramics, can achieve higher precision than using Rogers material as the substrate, and has been widely used in some special fields. Integrated antennas based on fan-out packaging have relatively high requirements for packaging materials in terms of electrical properties, mechanical properties, and manufacturability. In making use of the planar area to solve the electromagnetic coupling interference between the antenna and the chip, the large planar size and the improvements in three-dimensional integration are presenting challenges.

In 2019, Michigan State University used 3D additive printing manufacturing technology to make it easier to manufacture more complex structures such as cavities on the substrate, thereby reducing the impact of substrate loss. Among this technique, they used low dielectric materials to serve as the substrate. In addition, other technologies such as ion implantation and artificial magnetic conductors have also been used to improve the performance of the package-integrated antennas [5–7].

Georgia Tech proved for the first time in 2017 that, with use of a common heterogeneous integrated glass as the substrate material combined with the transmission line structure, antennas operating at 28 and 39 GHz could be realized. Their study used a microstrip line edge-fed patch antenna. In this way, the thickness of the substrate material was 100 μm, the insertion loss of the transmission line structure was 0.2–0.3 dB/mm, and there was a relative working bandwidth of 6% at the working frequency of 39 GHz.

Silicon materials are not conducive to the energy emission of the antenna radiation field because of their high dielectric constant. However, silicon-based materials are compatible with traditional microelectronic silicon-based processes. Thus they have outstanding advantages in process compatibility. In 2012, Leti, a French research institution, adopted edge feeding on a silicon substrate and used a back cavity structure to study the characteristics of folded dipole antennas and their arrays. A single antenna element could cover the 57–66 GHz frequency band. After their processing technique, the sample had good impedance matching and a gain greater than 5 dBi. The 1×4 antenna array demonstrated the ability to perform beam steering within a range of up to ±60 degrees. Georgia Tech and Shanghai Jiao Tong University also used bulk silicon processing technology to fabricate a similar cavity structure on a silicon substrate to reduce the equivalent dielectric constant of the radiation area and improve the radiation performance of the antenna. In 2017, the Indian Institute of Technology used the defective ground structure (DGS) to improve the narrow-band characteristics of the patch antenna due to the high dielectric and high loss of silicon, and the bandwidth was increased by four to eight times the original foundation. Brest University and The 14th Institute of CETC (China Electronics Technology Group Corporation) also used silicon substrates to implement chip-level dipole antennas. Dipole antennas are used in high-frequency bands such as Ka and W, which means they have advantages such as small size, large bandwidth, high rate, and other characteristics, and they can be utilized in high-performance phased array antennas, millimeter-wave image processing, and high-performance chip group communications. Table 7.1 shows the research status of the AiP.

As a potential competitor of three-dimensional radio-frequency heterogeneous integration, HR-Si interposer technology becomes more attractive when antennas move toward a higher working frequency and thus require a higher integration density. This chapter focuses on the research of our team in stacking integrated antennas based on a HR-Si interposer.

Table 7.1 Research status of AiP integrated antenna.

		ASE [8]	University of Florida [9]	INTEL [10]	GIT [11]	MSU [12]	CAS [13]	IBM [14]	Toshiba [15]
Date		2019.5	2019.5	2019.5	2019.5	2019.5	2019.5	2018.5	2018.5
Operating frequency (GHz)		26–33/7	67–80.88/13.88	25–30 and 35–40	0.2–1.4/1.2	10–25/15	24.8–31.3/6.5	90–98/8	2.39–2.45/0.065
Substrate material	Name	–	Rogers TMM4	PCB	PET	Multipoint jet printing aerosol	Rogers 6006	Si1-xGex	Epoxy resin
	ε_r	3.6	4.5	4.2–4.7	3.9	–	6.15	11.7	–
Size		$2.5 \times 2.5 \, mm^2$	$0.19 \, mm^2$	$50 \times 32.5 \times 1.5 \, mm^3$	$30 \times 40 \times 0.135 \, mm^3$	–	$2.1 \times 0.75 \times 2.8 \, mm^3$	$13.5 \times 11.3 \, mm^2$	$5.25 \times 9.0 \times 1.0 \, mm^3$
Antenna form		Patch	Inverted F	Patch	Patch	Vivaldi	Patch	Patch	Gap
Gain		10.3 dBi	1.77 dBi	–	–	12.5 dBi	10.1 dBi	2 dBi	−5.5 dBi

7.2 Theoretical basis of patch antenna

Compared with other devices in the radio-frequency system, the main problem of antennas is high-performance integration, as well as specific performance indicators (including size, pattern, gain, and bandwidth, etc.) that meet the established requirements. Currently, most of the types of antennas that can be integrated are patch antennas, slot antennas, dipole antennas, etc. The patch antenna has a simple structure and can be processed using existing microelectronic processes, which makes it one of the focuses of the industry.

The microstrip patch antenna is mainly composed of a substrate and two metal conductors on the upper and lower surfaces. The length and width of the patch are L and W, respectively, and the thickness of the substrate is h. Generally, the length L of the patch antenna takes half of the wavelength $\lambda/2$, where λ is the wavelength in the substrate. In transmission line model, patch antenna can be regarded as a microstrip line with two open-circuited ends. Due to the open-circuit ends and the half wavelength length L, a standing wave will be formed along the length direction as shown in Fig. 7.1A. Fig. 7.1B shows the electric field distribution in the side view of the path antenna. The electric field of the two patch fringes can be decomposed into horizontal and vertical components. The vertical electric components of the two fringes have same magnitude but are out-of phase, thus counteracting each other. While the horizontal electric components of the two fringes have same magnitude and same phase, thus reinforcing each other. Finally, the radiation of the patch antenna can be equivalent to the radiation of two slot antennas in parallel with the length direction.

There are two conventional feeding methods for patch antennas, edge feeding and coaxial feeding. Microstrip line and coaxial cable are used for feeding, respectively. The feeding position affects the matching of the antenna and the radiation characteristics of the antenna.

The main purpose of theoretical analysis of microstrip antennas is to predict the antenna's radiation performance, such as gain and directivity, and reduce the overall cost of engineering design. There are many theoretical analysis methods, and different methods have their own advantages, disadvantages, and limitations. The various parameters of the antenna have a mutual relationship with and influence on each other. Therefore choosing a suitable theoretical analysis method can reduce the cost of trial and error and improve efficiency. Table 7.2 shows the comparison of simulation analysis methods for different microstrip antennas [16–18]. In this chapter, numerical analysis methods based on finite element theory are used to realize the theoretical analysis of microstrip antennas.

7.3 Design of a patch antenna in stacked HR-Si interposers

Generally speaking, the substrate of the patch antenna should be made of a material with a low dielectric constant to increase the radiation efficiency of the antenna and allow more energy to be radiated into the space. However, the size of the patch antenna is also closely

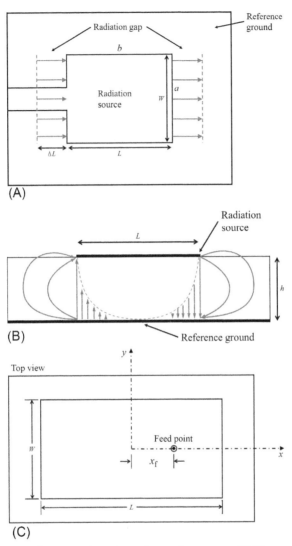

Fig. 7.1 (A) The top view of rectangular microstrip patch antenna, (B) The cross sectional view of rectangular microstrip patch antenna, and (C) The diagram of the coaxial-fed patch antenna.

related to the dielectric constant of the substrate, and a substrate material with a high dielectric constant is beneficial to reduce the size of the patch antenna. The back cavity structure of the patch antenna is easy to manufacture and integrate. Studies have shown that it can be used to improve the narrowband characteristics of the patch antenna [19,20]. The dielectric constant of the HR-Si substrate is higher than that of the commonly used materials for microwave printed circuit boards, and the processability of HR-Si substrate is good. In this section, the microcavity structure is introduced into

Table 7.2 Comparison of simulation analysis methods for different microstrip antennas.

Name	Principle	Advantages	Disadvantages
Vector potential function method	The rigorous solution of radiation can be obtained based on the field generated by the horizontal electric dipole determined by the wave propagation theory proposed by Sommerfeld, combined with numerical methods	The solution obtained is a strict solution of the radiation field	It is difficult to obtain rigorously solved explicit expressions, and the physical explanation is insufficient, which is not suitable for engineering applications
Dyadic Green's function method	Based on the double integral dyadic Green's function, the current density components J_x and J_y are obtained by considering the patch antenna as a two-dimensional transmission line, combining boundary conditions, Maxwell equations, and continuity equation	The solution obtained is a strict solution of the radiation field	Similar to the vector potential function method, the calculation is complicated and not easy to analyze
Traverse network model method	The radiating mechanism is regarded as a wire network composed of thin wires, and the wire segment current obtained by Richmond's reaction theorem is applied to obtain the antenna characteristics	When the wire network is finely divided, good results can be obtained	Considerable computing resources and storage resources are required, which increases the cost of engineering applications
Transmission line model method	In the vertical direction, the microstrip patch and the ground plane are regarded as a standing wave microstrip line with two open-circuited ends. The radiation of an open-circuited end can be equivalent to the radiation created by a slot. And then, radiation admittance and resonance frequency can be obtained by figuring out the equivalent magnetic current	Suitable for engineering applications, the corresponding physical model is simple and intuitive	It is only suitable for rectangular microstrip antennas. Besides the resonance point, the curve of input impedance versus frequency is not accurate; the admittance calculation is not accurate

Table 7.2 Comparison of simulation analysis methods for different microstrip antennas—cont'd

Name	Principle	Advantages	Disadvantages
Cavity model	Based on the analysis of the microstrip resonant cavity, according to the boundary conditions of the cavity, parameters such as resonant frequency, quality factor, and impedance are calculated from a fundamental mode in the cavity	The calculation cost and accuracy are suitable for engineering applications; multimode theory is developed on the basis of single-mode theory, which can more accurately reflect the field in the cavity	Does not consider the field change in the vertical direction; only suitable for calculating when the substrate thickness is much smaller than the wavelength in the substrate
Numerical analysis	The integral equation method is used to solve the field created by a point source under boundary conditions, and the principle of superposition is used to obtain the strict analytical solution of the total field, and the tedious calculus equation is replaced by the form of finite-difference summation to obtain sufficiently accurate numerical solutions	It has good flexibility and can effectively solve irregular patches and structures; the results are suitable for engineering applications with high accuracy	The computational cost is relatively large; the calculation result of the finite element method becomes worse with the increase of the fineness of the mesh; the error of solving the problem of the infinite region is relatively large

the substrate of the HR-Si TSV interposer, and the power supply is realized by TSV interconnection, as shown in Fig. 7.2; we can combine the advantages of the high dielectric constant of the silicon substrate, while minimizing the radiation effect of the high dielectric constant substrate material on the antenna. The on-chip integrated antenna can be bonded to another HR-Si TSV interposer through microsolder joints to achieve higher-density three-dimensional stacking integration. The bottom TSV interposer can integrate filters, power amplifiers, and other antenna feed network structures. The active and passive components can realize the three-dimensional integration of the system without increasing the area. Compared with the existing planar integration and fan-out integration, a higher degree of integration can be achieved [21].

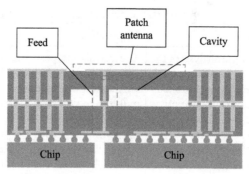

Fig. 7.2 Schematic diagram of the structure of the patch antenna integrated on the interposer.

In the design of the patch antenna, the length and width of the patch and the feeding position of the antenna are key design parameters. The width of the patch antenna can be calculated by Eq. (7.1).

$$W = \frac{c}{2f_0} \times \left(\frac{\varepsilon_r + 1}{2}\right)^{-\frac{1}{2}} \tag{7.1}$$

Where c is the speed of light in free space, f_0 is the operating frequency of the antenna and ε_r is the relative dielectric constant.

The calculation of the patch length will be a bit complicated, given by Eq. (7.2).

$$L = L_e - 2\Delta L \tag{7.2}$$

Where L_e is the effective length given by Eq. (7.3) and ΔL is the length of the equivalent slot antennas given by Eq. (7.4)

$$L_e = \frac{\lambda_e}{2} = \frac{\lambda_0}{2\sqrt{\varepsilon_e}} = \frac{c}{2f_0\sqrt{\varepsilon_e}} \tag{7.3}$$

$$\Delta L = 0.412h \frac{(\varepsilon_e + 0.3)\left(\frac{W}{h} + 0.264\right)}{(\varepsilon_e - 0.258)\left(\frac{W}{h} + 0.8\right)} \tag{7.4}$$

Where the λ_e is the effective wavelength, λ_0 is the free-space wavelength, and ε_e is the effective dielectric constant, given by Eq. (7.5).

$$\varepsilon_e = \frac{\varepsilon_r + 1}{2} + \frac{\varepsilon_r - 1}{2}\left(1 + 12\frac{h}{W}\right)^{-\frac{1}{2}} \tag{7.5}$$

Feeding position can affect the impedance matching of the antenna, so it should be carefully designed. The diagram of the coaxial-fed patch antenna is shown in Fig. 7.1C.

The fundamental mode of the patch antenna is TM_{10} and its magnitude of electric field does not vary along the width direction, so in theory, it can be excited in any point along the Wide direction. In order to avoid exciting high-order modes TM_{1n}, the feeding point generally takes in the center of the width side. As for the Length direction, the center is voltage node and its impedance is zero. The more offset from the center along the Length direction, the larger impedance can be obtained. As for coaxial probe feeding, the impedance is generally designed to be 50 Ω. X_f is the position of the feeding point along the Length direction, given by Eq. (7.6).

$$X_f = \frac{L}{2\sqrt{\xi_{re}(L)}} \tag{7.6}$$

Where $\xi_{re}(L)$ is given by Eq. (7.7).

$$\xi_{re}(L) = \frac{\varepsilon_r + 1}{2} + \frac{\varepsilon_r - 1}{2}\left(1 + 12\frac{h}{L}\right)^{-\frac{1}{2}} \tag{7.7}$$

After using these equations to obtain the corresponding preliminary design parameters, the relevant design parameters of the patch antenna are adjusted in combination with the simulation structure in HFSS, and finally the appropriate design parameters are obtained.

In order to verify the proposed design concept of the patch antenna with the back cavity in the stacked HR-Si interposer, a patch antenna is designed on a stacked HR-Si TSV interposer that works in the Q-band. It consists of upper and lower substrates, where the upper substrate includes a patch, a feeding TSV, and a back cavity, and the lower substrate mainly includes the CPW feed structure of the RDL layer.

The patch antenna feed line includes a TSV interconnection series structure in which two sections of the upper and lower interposers are connected end-to-end by microbump bonding and the CPW implementation on the RDL. In order to increase the probability of subsequent successful bonding, tin is used for the microbumps, with an outer diameter 100 μm larger than the diameter of the TSV. In order to verify the feasibility of the antenna feed structure, a daisy chain test structure [22] which is similar to the proposed feeding structure is designed and analyzed, and the simulated results show that this structure can operate at up to 40 GHz. It consists of microbumps [23], CPW transmission lines, and RF TSVs mutually connected in series, as shown in Figs. 7.3 and 7.4. Fig. 7.5 is the simulated S-parameters result of the test structure. It can be found that the overall insertion loss of the structure is 0.55 dB at 40 GHz. Using the deembed of HFSS, the transmission structure insertion loss corresponding to CPW1 and CPW3 can be obtained as 0.37 dB at 40 GHz, and CPW2 corresponds to the insertion loss of the transmission structure, which is 0.15 dB at 40 GHz, and the insertion loss of the transmission structure corresponding to TSV1+Bump1+TSV2 and TSV4+Bump2+TSV3 is

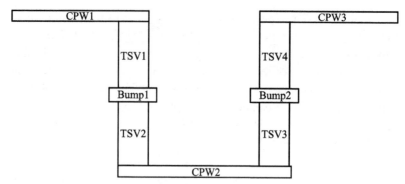

Fig. 7.3 Schematic diagram of test structure.

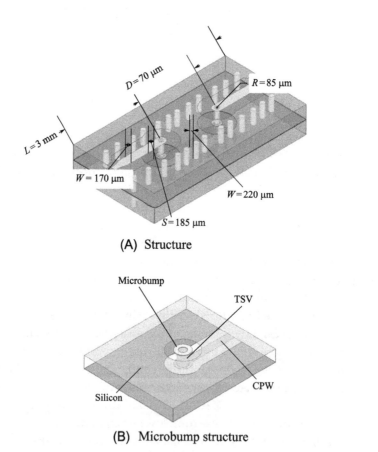

(A) Structure

(B) Microbump structure

Fig. 7.4 A test structure model composed of bumps, CPW transmission lines, and RF TSV interconnection in series.

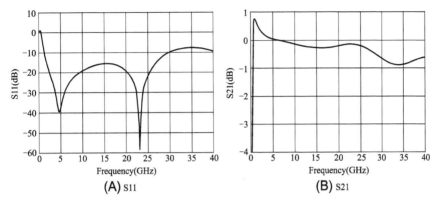

Fig. 7.5 Simulation results of the test structure.

0.26 dB at 40 GHz. Using the same process, the sample was fabricated, and the insertion loss of the stacked test structure composed of the TSV and microsolder joints in the feed structure achieved 1.7439 dB at 40 GHz.

Fig. 7.6 compares the S11 parameters and pattern of the patch antenna with and without cavity structure. It can be seen that, after the cavity structure is added to the substrate, the operating frequency of the antenna is increased by about 10 GHz, and the bandwidth changes from 140 to 1160 MHz; the radiation pattern and the maximum gain have also been improved. The maximum gain increased from about 1.4 to about 5.

On the basis of the preceding simulation analysis, this section continues the analysis of the influence of the main parameters, including the size of the cavity and the thickness of the upper and lower substrates. Figs. 7.7 and 7.8 show the influence on S11 and gain pattern after changing the height and size of the cavity, and the influence on S11 and gain pattern after changing the thickness of the upper and lower substrates, respectively.

Through parametric simulation analysis, it can be found that the size of the cavity will affect the radiation characteristic of the patch antenna. The working frequency of the

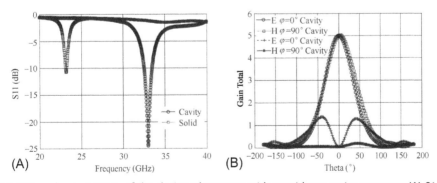

Fig. 7.6 S11 and gain pattern of the designed antenna with or without cavity structure: (A) S11 and (B) gain pattern.

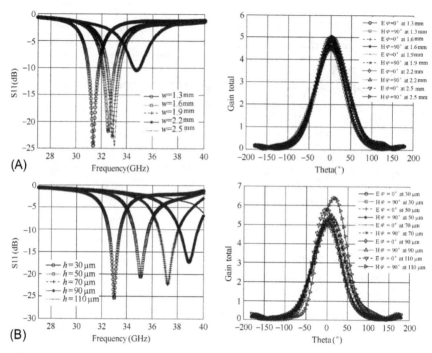

Fig. 7.7 The influence of the size and height of the cavity on the S11 and gain pattern of the antenna: (A) the size of the cavity and (B) the height of the cavity.

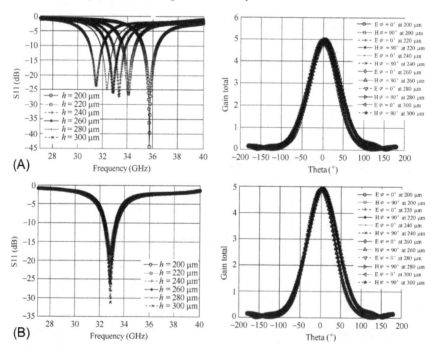

Fig. 7.8 The influence of the size of the substrate on the S11 and gain pattern of the antenna: (A) the thickness of the upper substrate and (B) the thickness of the lower substrate.

antenna increases as the size of the cavity increases, but the increase gets smaller and smaller. The value range of the cavity size in the study is 1.3–2.5 mm, and the working frequency of the antenna has been increased from 31.35 to 33 GHz. When the cavity size is greater than about 1.6 mm, the operating frequency of the antenna is basically stable at 32–33 GHz. The size of the cavity also has a certain impact on the bandwidth of the antenna. The bandwidth of the antenna has been increased from 290 to 960 MHz. As shown in Fig. 7.7A, as the size of the cavity gradually increases, the antenna's radiation and gain in the main radiation plane do not change significantly. After simulating the influence of the size of the cavity on the performance of the antenna, the influence of the height of the cavity on the performance of the antenna is also studied. The height of the cavity ranges from 0.03 to 0.11 mm. From the simulation results, we can conclude that as the height of the cavity gradually increases, the working frequency of the antenna gradually increases, and the increase is gradually reduced, from 33.5 to 38.94 GHz. It also has a significant impact on the working bandwidth of the antenna, from 390 to 1480 MHz. As for gain pattern, the height of the cavity has little impact on it, and the sudden change at 90 μm is quite different from other results, which is probably an exception.

The thickness of the upper and lower substrates will also affect the radiation performance of the patch antenna. The upper substrate is the radiation area of the patch antenna. As the thickness of the substrate increases, the operating frequency of the antenna is decreased, from 35.74 to 31.34 GHz. The bandwidth of the antenna does not change much, basically around 1 GHz. The gain does not change much at the minimum S11 frequency of each thickness; the radiation direction of the main lobe is at $\theta=0$ degrees, and the gain is about 5. The lower substrate is primarily related to the feed structure of the antenna, and mainly affects the electrical performance of the CPW transmission line structure. With the increase of the thickness of the lower substrate, the operating frequency of the antenna remains basically unchanged at about 33 GHz. The operating bandwidth of the antenna does not change much, basically around 1 GHz. The antenna pattern and gain do not change much, basically at the direction of $\theta=0$ degrees, and the gain is about 5.

If the millimeter-wave antenna is to be integrated into a heterogeneous integrated system, the key is how to minimize the impact of the patch antenna radiation on the chip integrated under it. Finally, combined with the actual process, the relevant parameters are determined as shown in Table 7.3, and the simulation model of the antenna in HFSS is shown in Fig. 7.9. Fig. 7.10 shows the simulated S11 parameter and gain patterns of the antenna. The minimum return loss S11 is −24 dB at 33.08 GHz, the bandwidth of −10 dB is 32.6–33.58 GHz, and the bandwidth is 0.98 GHz. The simulation result is that the maximum gain is 5 and the direction is $\theta=0$ degrees. Without the cavity-backed structure, the bandwidth of the antenna is 0.13 GHz at 23 GHz, the maximum gain is 3.5, and the direction is $\theta=0$ degrees. In HFSS, the distribution of the magnetic field of the upper and lower substrates of the antenna is shown in Fig. 7.11.

Table 7.3 Design parameters of patch antenna.

Parameter		Value
Thickness of substrate		250 μm
Patch	Width	2.3 mm
	Length	1.7 mm
Back-cavity	Width	5 mm
	Length	5 mm
	Height	30 μm
TSV	Diameter	200 μm
Microbump	Inside diameter	300 μm
	Outside diameter	220 μm

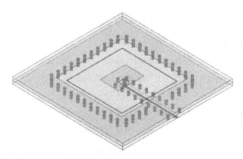

Fig. 7.9 Simulation model of antenna in HFSS.

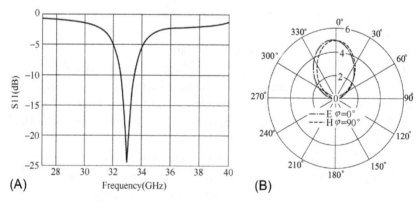

Fig. 7.10 Antenna S11, gain pattern simulation results: (A) S11 and (B) gain pattern.

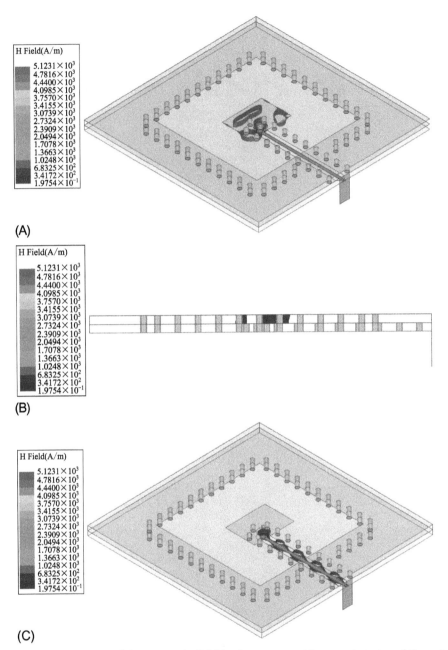

Fig. 7.11 The distribution of the magnetic field in the upper and lower substrates of the antenna (A–G). (A) Perspective view of the distribution of the magnetic field in the upper substrate with the cavity-free structure. (B) Side view of the distribution of the magnetic field in the upper substrate with the cavity-free structure. (C) Perspective view of the distribution of the magnetic field in the lower substrate with the cavity-free structure.

(Continued)

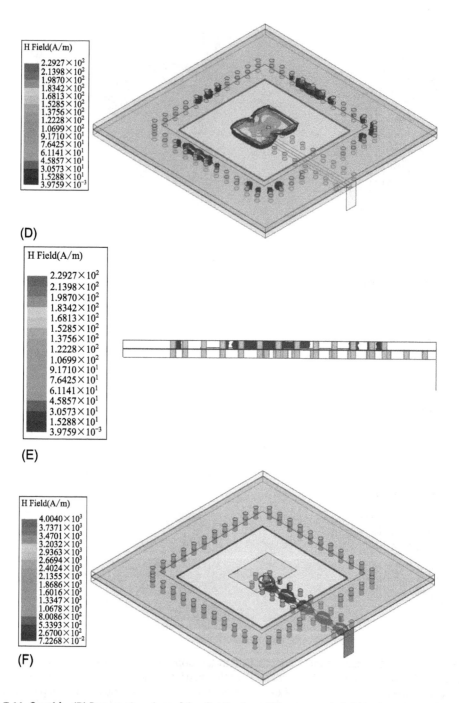

Fig. 7.11, Cont'd (D) Perspective view of the distribution of the magnetic field in the upper substrate with the cavity structure. (E) Side view of the distribution of the magnetic field in the upper substrate with the cavity structure. (F) Perspective view of the distribution of the magnetic field in the lower substrate with the cavity structure.

(Continued)

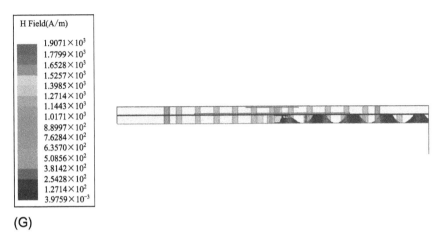

Fig. 7.11, Cont'd (G) Side view of the distribution of the magnetic field in the lower substrate with the cavity structure.

It can be seen that:
(1) After a cavity is etched in the antenna, the intensity of the magnetic field in the cavity area increases.
(2) Because a large grounding plane is added between the upper and lower substrates, the lower layer area is not affected by the strong radiation interference of the upper antenna.

7.4 Processing of a patch antenna in stacked HR-Si interposers

The patch antenna integrated in the stacked HR-Si interposers is formed by bonding two HR-Si interposers. The process flow of the HR-Si interposer is the same as described in Chapters 2 and 3. The overall process flow is shown in Fig. 7.12. The process parameters are shown in Table 7.4.

Based on the aforementioned HR-Si TSV interposer process flow, the proposed antenna is processed. Fig. 7.13 shows the optical photo of the microbump after removing the seed layer. Fig. 7.14 is an X-ray inspection photo of the partial structure of the antenna after removing the seed layer. Fig. 7.15 is a photo of the upper and lower substrates before bonding. Fig. 7.16 is a photo of a diced antenna sample after bonding. The outline size of the antenna is 10mm × 10mm without optimization.

7.5 Test and analysis of patch antenna in stacked HR-Si TSV interposer

In the experiment, an Agilent vector network analyzer [24], Cascade Microtech Infinity probe, and Cascade Microtech Summit 12,000 semiautomatic probe station were used to

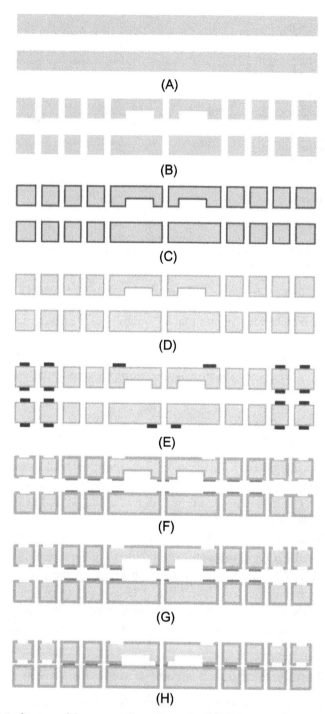

Fig. 7.12 Schematic diagram of the process flow of a stacked HR-Si TSV interposers integrated antenna with an embedded back cavity.

Table 7.4 Main process parameters of patch antenna in stacked HR-Si interposers.

Parameter		Value
Resistivity of substrate		1400–4000 Ω·cm
TSV	Diameter	200 μm
	Thickness of Cu layer	10 μm
	Thickness of SiO$_2$ layer	0.1 μm
RDL	Thickness of Cu layer	10 μm
	Thickness of SiO$_2$ layer	0.1 μm
	Thickness of microbump	5 μm

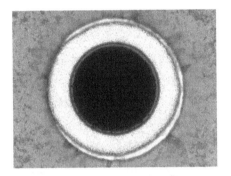

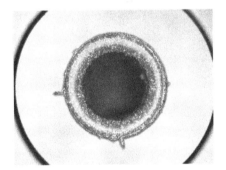 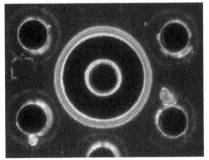

Fig. 7.13 Photos of the microbump structure of the antenna.

measure test structure samples firstly. During the test, since the bottom and the periphery of the sample contained a large number of patterned metal structures, it was necessary to ensure that the bottom of the sample was insulated from the metal chuck of the probe station to prevent short circuits. The test procedure is as follows. First, carry out the DC resistance test, and screen out the test structure with transmission function. Then S-parameter measurement is performed. Fig. 7.17 shows the test result of a test structure under 0.01–40 GHz with a scanning step size of 0.01 GHz at room temperature. The total

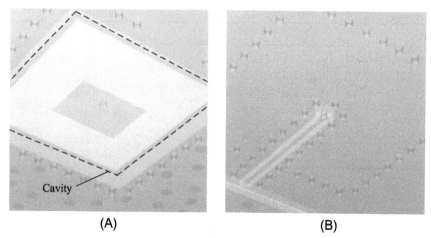

Fig. 7.14 X-ray photos of antenna sample: (A) upper substrate and (B) lower substrate.

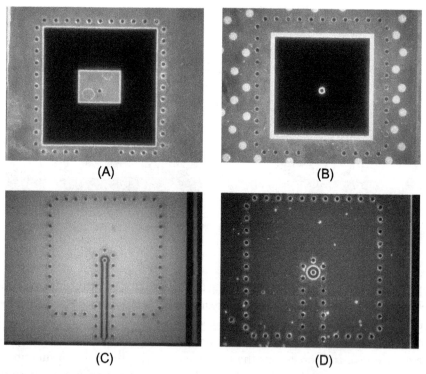

Fig. 7.15 Unbonded samples: (A) upper substrate with metal patch; (B) upper substrate with back cavity; (C) lower substrate with CPW line; and (D) lower substrate with microbumps.

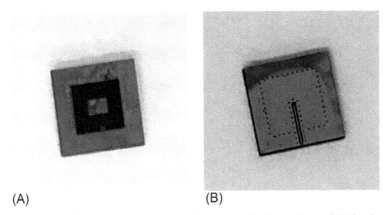

Fig. 7.16 Photographs of the antenna samples after dicing: (A) front side and (B) back side.

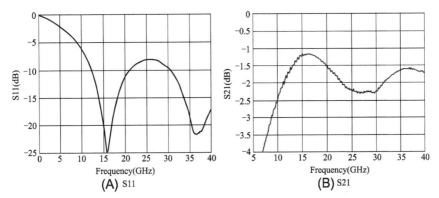

Fig. 7.17 Test result of a single test structure under the conditions of 0.01–40 GHz and 0.01 GHz step size at room temperature.

length of the test structure is 3.71 mm, including microbumps, TSV, and CPW. From the test results, the insertion loss of the test structure is 1.7439 dB at 40 GHz.

Before the antenna is bonded, the S11 parameters of the bare antenna chip is tested. The test result is shown in Fig. 7.18. The resonance point is at 32.76 GHz with −25.0 dB, the simulated value of S11 is 33.08 GHz with −24.4 dB, the deviation between experiment and simulation is 0.32 GHz which implies a relative error of 1.05%. The measured −10 dB bandwidth is 1.46 GHz ranging from 32.14 GHz to 33.60 GHz. Compared with the simulated value of 0.98 GHz, the bandwidth is 0.48 GHz larger.

Next, the test fixture is designed and processed, and the assembly is carried out. Rogers 5880 substrate was used to make CPW transmission line to install SMA connectors, and then the antenna was assembled on the Rogers substrate and gold wire was used to realize the electrical connection between the two, as shown in Fig. 7.19.

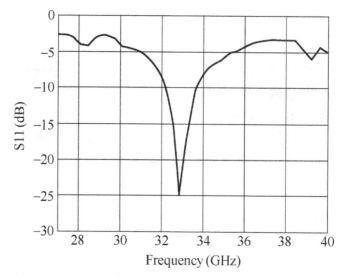

Fig. 7.18 Test result of antenna bare chip.

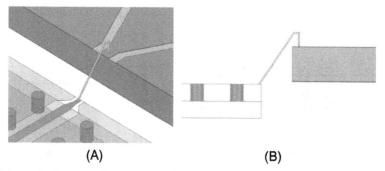

Fig. 7.19 Schematic diagram of antenna test fixture assembly and connection: (A) partial view and (B) side view.

Fig. 7.20 shows the test structure HFSS simulation analysis model, including SMA connector, and CPW transmission structure on the PCB. The simulation results are shown in Fig. 7.21. The assembled antenna sample is shown in Fig. 7.22.

After assembly, use the Agilent vector network analyzer to test the S11 parameters of the antenna and the measured result is shown in Fig. 7.24 A. The resonant frequency of the antenna is at 32.75 GHz with −16.3 dB. Compared with the simulated value of 33.08 GHz with −24.4 dB, the resonant frequency shifts 0.33 GHz lower and has a relative error of 1.0%. The measured −10 dB bandwidth of the assembled antenna is 1.04 GHz ranging from 32.23 GHz to 33.27 GHz. Compared with the simulated value of 0.98 GHz, the measured bandwidth is 0.06 GHz larger. It is obvious that the performance of S11 of the assembled antenna is well matched with the diced antenna before assembly.

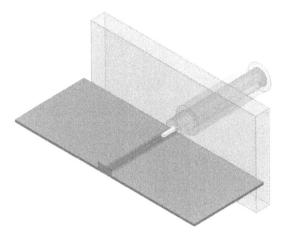

Fig. 7.20 Schematic diagram of simulation structure of antenna test fixture.

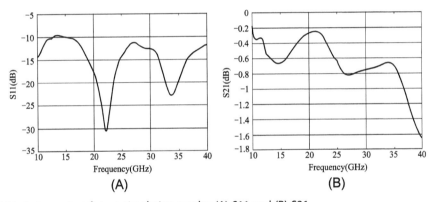

Fig. 7.21 Antenna test fixture simulation results: (A) S11 and (B) S21.

Fig. 7.22 Antenna assembly test diagram: (A) front side and (B) back side.

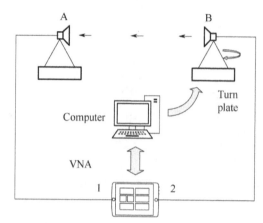

Fig. 7.23 Schematic diagram of antenna pattern test method.

The radiation pattern of the antenna was measured in an anechoic chamber. The antenna under test acts as a transmitting antenna and is mounted on a rotating platform which can rotate within the range of −90 to 90 degree, while a standard horn antenna acts as a receiving antenna and is mounted on another platform in the opposite direction as shown in Fig. 7.23. The two ports of the vector network are respectively connected to the standard horn antenna and the antenna under test. The standard horn antenna receives the signals radiated by the antenna under test at different angles and passes the data to VNA. The signal intensity is processed to obtain the gain of the proposed antenna in different angles. The measured maximum gain is 3.0 and the main lobe is roughly at $\theta=0$ degree as shown in Fig. 7.24 B. The main lobes of the E-plane and H-plane are split, and there are large side lobes. Table 7.5 gives the researches of integrated antennas on silicon substrate or silicon interposer.

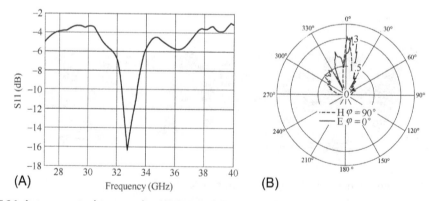

Fig. 7.24 Antenna actual test results: (A) S11 and (B) gain and direction diagram.

Table 7.5 Comparison of related research results.

Publication	Date	Working frequency/bandwidth	Size	Antenna type	Gain
CEA-Leti [25]	2012.6	60 GHz/9 GHz	$2.3 \times 2.2\,mm^2$	Folded dipole antenna	7.9 dBi (6.17)
SJU [26]	2019.5	3.47 GHz/0.34 GHz and 6.5 GHz/1.24 GHz	—	Cavity-backed antenna	1.9 dBi (1.55)
Lab-STICC [27]	2018.5	35 GHz/11.1 GHz	$10 \times 10 \times 0.655\,mm^3$	Dipole antenna	1.69 dBi (1.48)
GIT [28]	2016.1	100 GHz/13.35 GHz	$1.1 \times 0.65\,mm^2$	Patch antenna	4.4 dBi (2.75)
Indian Institute of Technology [29]	2017.10	10.6 GHz/0.4 GHz	$4 \times 4.3\,mm^2$	Patch antenna	4.0 dBi (2.51)
CECT 55 [30]	2019.5	100 GHz/11.5 GHz	$2.93 \times 1.0\,mm^2$	Dipole antenna	6.03 dBi (4.01)
This work		32.75 GHz/1.04 GHz	$6 \times 6 \times 0.55\,mm^3$	Patch antenna	4.77 dBi (3)

Comparing the simulation results with the test results, it can be seen that there is a certain variation in the S11 parameters and the radiation pattern among the simulation results and test results. The resonant frequency shift about 0.05 GHz, the return loss is greater at 8.07 dB after assembly, and the bandwidth is smaller at 0.35 GHz. While for Gain pattern, the main lobes of the E-plane and H-plane are split, and there are large side lobes.

The deviation might mainly be attributed to the following two factors. One is the manufacturing errors and the other is that in order to test the radiation pattern of the proposed antenna, a gold wire is used to connect the antenna and the fixture, and this may introduce parasitic inductance and result in impedance mismatch. As the frequency increases, the parasitic inductance and parasitic resistance of the gold wire bonding structure increases. In addition, the gold wire and the surrounding PCB board and the metal structure on the HR-Si form a coupling capacitor. The bonding gold wire has an effect on the working frequency, bandwidth, and pattern of the patch antenna [31]. The bandwidth of the antenna generally decreases with the increase of the length of the bonding gold wire; the center frequency shifts to lower frequency as the length of the bonding gold wire increases. The gain will decrease as the length of the gold wire increases. The influence of the gold wire bonding process on the antenna can be effectively alleviated by replacing a single bonding gold wire to three bonding gold wires to form a GSG structure, and designing appropriate spacing and arch height can achieve the feeding of 60 GHz high-frequency antennas [32].

7.6 Summary

In order to explore the application of the HR-Si interposer to 5G and millimeter-wave antennas, this chapter reported on the design of a patch antenna in stacked HR-Si interposers with buried cavities to optimize the bandwidth and gain. The main factors affecting the performance of the integrated patch antenna were analyzed, and samples were produced, assembled, and verified. The test results show that it has a bandwidth of 1.04 GHz and a maximum gain of 3 when operating at 32.75 GHz. The error was also analyzed and a method for optimization was derived.

References

[1] Wang W, Qiu S, Wang J, et al. Research development of millimeter-wave antenna integration technology. Microelectronics 2019;049(4):551–7. 573. [in Chinese].
[2] Zhou X. Review of advanced packaging technology. Integr Circ Appl 2018;35(297(6)):8–14 [in Chinese].
[3] Luo R. Research on the technology of inlined silicon interposer for fan-out wafer level packaging. Xiamen: Xiamen University; 2018 [in Chinese].
[4] Yan J. Research of process and application of 3D RF integrated low loss TSV interposer. Xiamen: Xiamen University; 2018 [in Chinese].

[5] Zhang T. Design and implementation of millimeter wave encapsulated antenna. Nanjing: Southeast University; 2017 [in Chinese].
[6] Watanabe AO, Ali M, Tehrani B, et al. First demonstration of 28 GHz and 39 GHz transmission lines and antennas on glass substrates for 5G modules. In: 2017 IEEE 67th electronic components and technology conference (ECTC). IEEE; 2017.
[7] Lu Y, Fang B, Mi H, et al. Mm-wave antenna in package (AiP) design applied to 5th generation (5G) cellular user equipment using unbalanced substrate. In: Electronic components and technology conference; 2018.
[8] Hsieh R, Chu F, Ho C, et al. Advanced thin-profile fan-out with beamforming verification for 5G wideband antenna. In: Electronic components and technology conference; 2019.
[9] Hwangbo S, Bowrothu R, Kim H-i, et al. Integrated compact planar inverted-F antenna (PIFA) with a shorting Via Wall for millimeter-wave wireless chip-to-chip (C2C) communications in 3D-SiP. In: Electronic components and technology conference; 2019.
[10] Thai T, Dalmia S, Hagn J, et al. Novel multicore PCB and substrate solutions for ultra broadband dual polarized antennas for 5G millimeter wave covering 28GHz & 39GHz range. In: Electronic components and technology conference; 2019.
[11] Zhou Y, Sivapurapu S, Chen R, et al. Study of electrical and mechanical characteristics of inkjet-printed patch antenna under uniaxial and biaxial bending. In: Electronic components and technology conference; 2019.
[12] Gjokaj V, Crump C, Papapolymerou J, et al. Vivaldi antenna array fabricated using a hybrid process. In: Electronic components and technology conference; 2019.
[13] Xue M, Wan W, Wang Q, et al. Wideband low-profile Ka-band microstrip antenna with low cross polarization using asymmetry AMC structure. In: Electronic components and technology conference; 2019.
[14] Gu X, Liu D, Baks C, et al. An enhanced 64-element dual-polarization antenna array package for W-band communication and imaging applications. In: Electronic components and technology conference; 2018.
[15] Yamada K, Sano M, Higaki M, et al. Small shielded bluetooth module equipped with slot antenna on the surface. In: Electronic components and technology conference; 2018.
[16] Zeng G. Miniaturization design of MEMS microstrip antenna based on metamaterials. Chengdu: University of Electronic Science and Technology of China; 2017 [in Chinese].
[17] Yin J. Research on microstrip antenna broadband technology. Beijing: Beijing Jiaotong University; 2007 [in Chinese].
[18] Wang F. Research on miniaturized microstrip antenna. Nanjing: Nanjing University of Posts and Telecommunications; 2019 [in Chinese].
[19] Zhang T, Zhang Y, Yu S, et al. A Q-band dual-mode cavity-backed wideband patch antenna with independently controllable resonances. In: International symposium on antennas and propagation; 2013. p. 118–21.
[20] Cheng J, Dib N, Katehi LP, et al. Theoretical modeling of cavity-backed patch antennas using a hybrid technique. IEEE Trans Antennas Propag 1995;43(9):1003–13.
[21] Sun Y, Chai Y, Ma S, et al. Design and analysis of a stackable structure patch antenna based on interposer. Electron Com Mater 2018;37(Supp I):76 [in Chinese].
[22] Ebefors T, Fredlund J, Perttu D, et al. The development and evaluation of RF TSV for 3D IPD applications. In: IEEE international 3D systems integration conference; 2013. p. 1–8.
[23] Sun Y, Jin Y, et al. Design and fabrication of 3D interconnected transmission structure based on TSV. In: International conference on microwave and millimeter wave technology; 2020.
[24] Deng C, Tang M, Hu Y, et al. Application of RF measurement and control technology for agilent vector network analyzer E5072A. Electron Packag 2019;19(05):8–11 [in Chinese].
[25] Dussopt L, Lamy Y, Joblot S, et al. Silicon interposer with integrated antenna array for millimeter-wave short-range communications. In: International microwave symposium; 2012. p. 1–3.
[26] Sun Y, Sun Y, Luo J, et al. Enhancing efficiency of antenna-in-package (AiP) by through-silicon-interposer (TSI) with embedded air cavity and polyimide dielectric micro-substrate. In: Electronic components and technology conference; 2019.

[27] Masri IE, Le Gouguec T, Martin P, et al. Integrated dipole antennas and propagation channel on silicon in Ka band for WiNoC applications. In: IEEE 22nd workshop on signal and power integrity (SPI); 2018. p. 1–4.
[28] Thadesar PA, Bakir MS. Fabrication and characterization of polymer-enhanced TSVs, inductors, and antennas for mixed-signal silicon interposer platforms. IEEE Trans Compon Packag Manuf Technol 2016;6(3):455–63.
[29] Gupta N, Abegaonkar MP, Koul SK, et al. Antenna on silicon with bandwidth improvement. In: International symposium on antennas and propagation; 2017. p. 1–2.
[30] Deng Y, Zhu F, Li B, et al. Design of a compact W-band planar dipole antenna on a single silicon substrate. In: International conference on microwave and millimeter wave technology; 2019.
[31] Song Y. Research on microwave MMW on-chip/encapsulated antenna technology. Chengdu: University of Electronic Science and Technology of China; 2017 [in Chinese].
[32] Felic G, Thomas C, Skafidas E, et al. Design of co-planar waveguide-fed slot/patch antenna with wire bond for a 60-GHz complementary metal-oxide-semiconductor transceiver. IET Microwaves Antennas Propag 2011;5(4):490–4.

CHAPTER 8

Through glass via technology

8.1 Introduction

In the fifth-generation mobile communication (5G) era, the trend of radio-frequency (RF) integration towards high electrical performance, compact structure, and low manufacturing cost drives the development of various advanced package technologies. Due to its excellent electrical performance and low cost, glass is regarded as a promising material in 5G applications [1]. Through glass via (TGV) technology is critical to realization of three-dimensional (3D) integration and has potential applications in MEMS capping, 2.5D interposers, integrated passive devices (IPDs), biodevices, and optical device integration [2]. The formation of TGVs with high aspect ratio and metallization by fully filled electroplating are key challenging processes. In this chapter, TGV fabrication technology using laser-induced deep etching (LIDE) is introduced. The methods for metallization are reviewed and tried out. Based on TGV fabrication, integrated passive device (IPD) technology and the development of embedded glass fan-out (eGFO) package technology are presented.

8.2 TGV fabrication

In the past decades, TGV fabrication has been researched by industry and research institutions. Numerous methods have been reported for forming vias in glass by chemical or physical mechanisms, each with its own benefits and drawbacks. A comparison between different manufacturing methods of TGV is shown in Table 8.1 [3–6]. Due to the disadvantages of high cost, lack of precision, and damage to several via formation solutions, the LIDE solution was studied. As is shown in Fig. 8.1, this method is divided into two steps: (1) the glass is first irradiated by picosecond laser, the refractive index change or nanovoids formation occurs along the laser irradiation zone and (2) then the glass is selectively etched in hydrofluoric acid (HF) solution, due to a higher reaction rate in the laser irradiation zone in HF solution [7,8]. Studies have been carried out on TGV fabricated in quartz glass using the LIDE method. Such TGVs are featured with fine pitch and vertical sidewalls. However, the cost of quartz glass is relatively high and the time required for the wet etching is long. These are crucial problems to be solved before commercial large-scale application of LIDE.

Here, the LIDE method is examined and applied to fabricate vias in borosilicate glass (SCHOTT AF32® eco) with low cost, good electrical performance, and thermal

Table 8.1 Comparison between different manufacturing methods of TGV.

Methods	Benefits	Drawbacks
Sand-blasting	Simple process	Large aperture, large pitch
Photosensitive glass	Simple process, high density, and high aspect TGV	High cost, inconsistency in accuracy of different structures
Electrical discharging	High aspect TGV	Still no commercial equipment available
Plasma etching	Small roughness of sidewall, no damage	Complicated process, high cost, and low speed
Laser ablation	High density and high aspect ratio TGV	Large roughness of sidewall, and damage
Chemical	Low cost, simple equipment, high speed	Large aperture
LIDE	High speed, high density, high aspect ratio TGV and no damage	Not vertical enough and expensive equipment

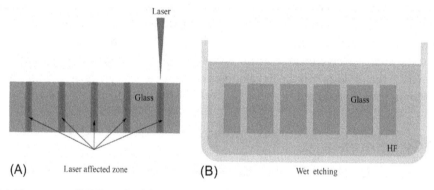

Fig. 8.1 Two steps of LIDE method, laser treatment and wet etching.

reliability. To disclose the via formation mechanism, glass samples are treated with single-pulse laser beam ($E = 55\,\mu J$/pulse and pulse width $\tau_p = 16\,ps$). The profile of the glass irradiated zone is inspected with SEM and shown in Fig. 8.2; a succession of nanovoids along the laser path can be found, where the nanovoids in the position of entry and exit are covered by fused glass. On the one hand, the HF solution enters the glass easier due to the nanovoids. On the other hand, the contact area between glass and the HF solution is enhanced by the nanostructure. Due to these two points, the reaction rate between the laser-irradiated region and HF is much higher than that between the nonirradiated region and HF, leading to the phenomenon of selective etching. When the glass after laser irradiation is immersed in the HF solution, there will be a taper angle in the via due to the difference between the etching rate along the irradiated path and the perpendicular direction of the path.

Based on the phenomenon of selective etching, the profile of the via can be optimized by the concentration of the HF solution. The glass is immersed in the HF solution at

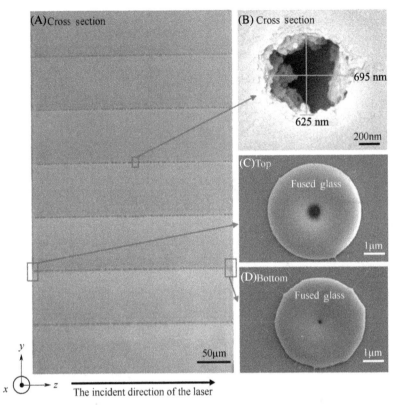

Fig. 8.2 (A) A series of nanovoids along the laser-irradiated path, (B) cross section of nanovoids in the middle path, (C) the nanovoids in the position of entry covered by fused glass, and (D) the nanovoids in the position of exit covered by fused glass.

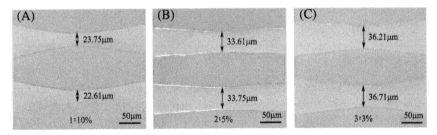

Fig. 8.3 The via profile of three glass samples in 10%, 5%, and 3% HF solution.

concentrations of 10%, 5%, and 3% to form TGV with a designed diameter of 60 μm. As shown in Fig. 8.3, the taper angles of the three samples are measured at 80.65, 83.07, and 84.18 degrees. The result shows that the profile of TGV can be effectively improved by decreasing the concentration of the HF solution.

The LIDE technology was also applied to another borosilicate glass SCHOTT Borfloat® 33. Table 8.2 lists the chemical composition of the two kinds of glass, and

Table 8.2 The chemical composition of two kinds of glass.

Glass type	Chemical composition	Single-shot laser beam parameters	Via sidewall angle (θ) (degrees)
SCHOTT AF32® eco	SiO_2 (60%–70%) B_2O_3 (10%–20%) Al_2O_3 (10%–20%) Others (1%–10%)	$E = 55\,\mu J/pulse$ $\tau_p = 16\,ps$	80.65
SCHOTT Borofloat® 33	SiO_2 (81%) B_2O_3 (13%) Al_2O_3 (2%) Others (4%)	$E = 45\,\mu J/pulse$ $\tau_p = 16\,ps$	86.53

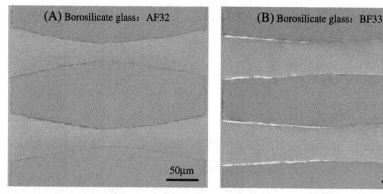

Fig. 8.4 The profile of TGV formed on two kinds of glass: (A) AF 32 and (B) BF 33.

Fig. 8.4 shows the profile of TGV formed on the two kinds of glass. The taper angle of BF33 is more vertical than that of AF32, due to the higher proportion of SiO_2.

8.3 Metallization of TGV

In addition to via formation in glass, high-quality metallization of TGV is another challenge. On the one hand, the via aperture of glass is much larger than TSV, which increases the duration of the copper electrochemical deposition (ECD) process. On the other hand, the adhesion between glass and common metals like copper is not as strong as that of silicon, which could result in a risk of peeling.

The first solution to realizing fully filled TGV is bottom-up electroplating TSV technology [9]. It starts with blind via formation. Then the seed layer is deposited by physical vapor deposition (PVD). Fig. 8.5 illustrates that the blind via is fully filled by a bottom-up copper electroplating process from Sky-Semi, where TGV with aspect ratio of 3:1 is deposited with a seed layer of 200-nm titanium and 1000-nm copper.

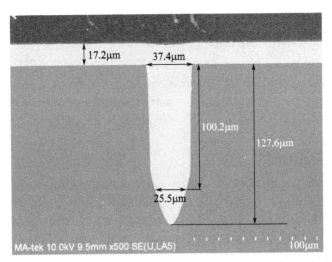

Fig. 8.5 The blind via is fully filled by bottom-up electroplating process from Sky-Semi.

Another metallization method is similar to that for the HR-Si interposer described in Chapter 2. First, TGVs are formed on the glass wafer, and then layers of 200-nm Ti and 1000-nm Cu are deposited conformally by double side sputtering sequentially, as the barrier layer and seed layer. Then dry films are laminated on the double sides and the photolithography is completed to realize double-sided patterns. Finally, the double-sided copper electroplating process is finished to metallize TGV and form RDL on both sides at the same time. The wafer is fixed on the hanger only along the edge; spraying or stirring can only be used during the process to enhance the exchange ability of the copper plating solution. In order to ensure the consistency of double-sided current density when the wafer is energized, a dual power supply is used. In the meantime, the plating areas on both sides are different, so it is a challenge to ensure the consistency of copper deposition on both sides. Design and optimization are needed to balance the plating area on double sides. Fig. 8.6 shows a partially filled and fully filled TGV sample of double-sided Cu electroplating.

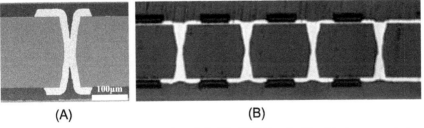

(A) (B)

Fig. 8.6 (A) Partially filled TGV sample and (B) fully filled TGV by double-sided copper electroplating process.

8.4 Passive devices based on TGV technology

8.4.1 Technology description

In RF front-end systems, passive devices such as matching devices, filters, multiplexers, couplers, transformers, baluns, and antennas are used broadly, as shown in Fig. 8.7. Due to the low loss property of glass, a high quality factor can be achieved for these passive devices. Based on TGV technology, 3D integration with high density can be realized.

8.4.2 MIM capacitor

Capacitance is essential for RF circuits that play the roles of DC isolation, filtering, coupling, tuning, rectifier, and so on. Usually, on-chip capacitors consist of microstrip capacitors, interdigital capacitors, and metal–insulator–metal (MIM) capacitors. MIM capacitors are widely used in integrated passive components (IPD). Fig. 8.8 illustrates the cross-section of a MIM capacitor on a glass substrate in simulation. The layers of electrode, passivation, top plate, dielectric, bottom plate, and glass stack from top to bottom. The basic parameters used in the simulation are listed in Table 8.3.

Fig. 8.9 illustrates the impact of the thickness of the top plate and bottom plate on capacitance and quality factor. With the thickness of the bottom plate increasing from 0.15 to 20 μm, the capacitance increases but quality factor decreases. The impact of the thickness of the top plate is consistent with that of the bottom plate. The edge effect will occur with the increase of the thickness of the capacitor plate, leading to the enhancement of the edge scattering electric field of the MIM capacitor, which in turn increases the parasitic capacitance and finally leads to an increase in the whole capacitance value. On the other hand, with the enhancement of the edge scattering electric field, the energy lost will also increase, leading to a decrease of the quality factor. Therefore it is necessary to decrease the thickness of the plate to improve the electrical performance of the capacitor.

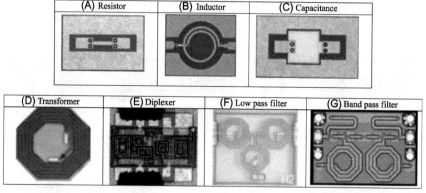

Fig. 8.7 Integrated passive devices.

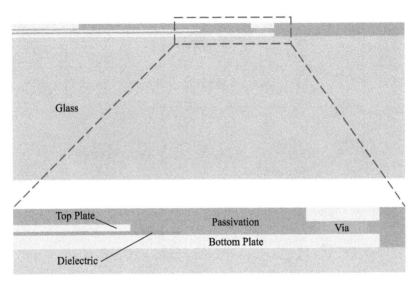

Fig. 8.8 The cross section of MIM capacitor simulation model.

Table 8.3 Parameters of MIM capacitor in simulation.

Description		Symbol	Value
Substrate (glass)	Thickness	H	230 μm
	Relative dielectric constant/loss tangent of glass	$\varepsilon_r/\tan\delta$	5.1/0.0035 at 1 GHz
Top plate	Thickness	h_{top}	3 μm
Dielectric (Si_3N_4)	Thickness	T	200 nm
	Relative dielectric constant of Si_3N_4	ε_{r1}	6
Bottom plate	Thickness	h_{bottom}	7 μm
Overlap area of capacitor		S	749.5 μm × 749.5 μm

Fig. 8.10A shows a schematic view of the MIM capacitor, including adhesive layers to improve the adhesion between copper and Si_3N_4. Fig. 8.10B and C illustrate the profile and top view of a fabricated MIM capacitor. The capacitors listed in Table 8.4 are fabricated and measured by vector network analyzer (VNA) Agilent N5244A. According to the comparison in Fig. 8.11, the error value between the simulation results and the test results of the MIM capacitor is kept within 10%, which proves the feasibility of the MIM capacitor process on glass.

8.4.3 TGV-based bandpass filter

Based on the 3D TGV inductor and MIM capacitor technology, a bandpass filter with properties of high electrical performance, compact size, and low cost can be realized in

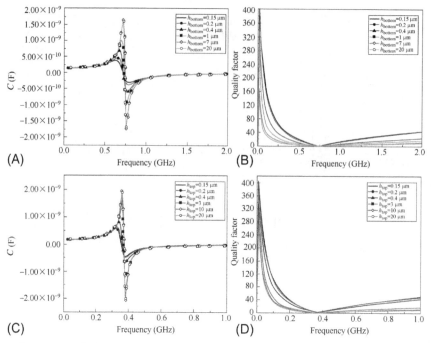

Fig. 8.9 The impact of thickness of bottom plate and thickness of top plate on capacitance and quality value.

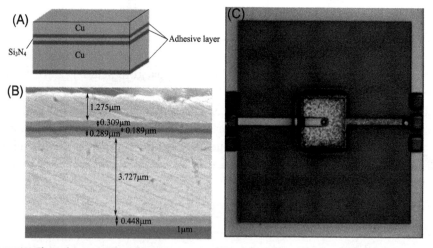

Fig. 8.10 (A) The schematic view of MIM capacitor, (B) the profile of a fabricated MIM capacitor, and (C) the top view of a fabricated MIM capacitor.

Table 8.4 Consistency of design values and measured values of glass-based MIM capacitors.

Groups	Top plate size (μm × μm)	Designed value (pF)	Measured value (pF)	Resonant frequency (GHz)	Error value
1	282.9 × 282.9	40.1 at 0.6 GHz	36.3 at 0.6 GHz	0.9	9.5%
2	331.7 × 331.7	72.1 at 0.6 GHz	71.9 at 0.6 GHz	0.78	0.3%
3	612 × 612	109 at 0.1 GHz	104 at 0.1 GHz	0.45	4.6%
4	749.5 × 749.5	165 at 0.1 GHz	159 at 0.1 GHz	0.36	3.6%
5	976.4 × 976.4	290 at 0.1 GHz	290 at 0.1 GHz	0.3	0

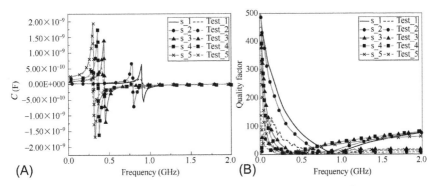

Fig. 8.11 The comparison between simulation result and test result of (A) capacitance and (B) quality factor.

mass production [10,11]. As shown in Fig. 8.12, the filters implemented on LTCC, laminate, silicon, acoustic wave, and 3D glass substrate are compared in terms of performance, thickness, size, density, frequency, and cost [1]. The acoustic wave filter is applied in 4G and LTE networks due to high selectivity but faces challenges when it reaches high frequencies. A 3D glass filter is a good choice in the high-frequency domain.

Fig. 8.13 shows the process of a TGV-based bandpass filter. The process begins with the TGV formation on a 6-in. 230-μm glass wafer. The diameter of the TGV is 60 μm and the taper tangle is 81.9 degrees. Then double-sided PVD is applied to form the seed layer with a thickness of 200-nm titanium and 1000-nm copper. After double-sided PVD, the lithography is performed on both sides of the wafer to define the patterns. Then the double-sided copper electroplating is finished to fully fill the TGV and to realize RDL on both surfaces of the wafer at the same time. At this point, the 3D TGV inductor and bottom plate of the capacitor are completed. The MIM capacitor is fabricated by

	LTCC	Laminate	Silicon	Acoustic Wave	3D Glass
Performance	Low loss	Substrate loss	Substrate loss	High selectivity	Low loss
Thickness	Thick	Thin	Thin	Thin	Thin
Size	Small footprint	Moderate footprint	Moderate footprint	Small footprint	Small footprint
Density	Low density	Moderate density	Moderate density	High density	High density
Frequency	Low–high	Low–moderate	Low–moderate	Low	Low–high
Cost	Mass production	Less in demand	High	High	Not in market

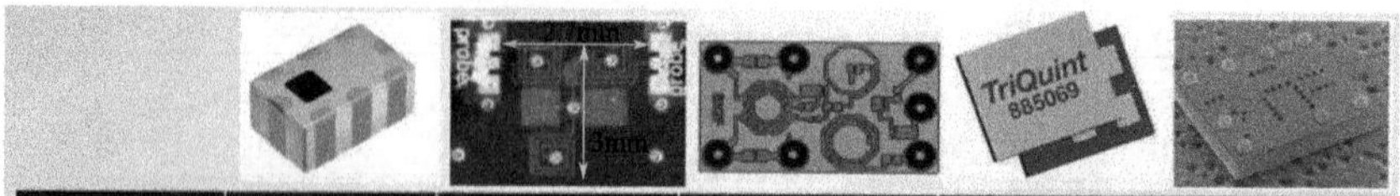

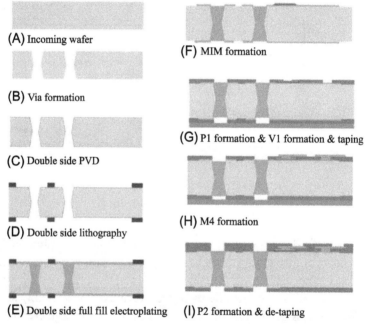

Fig. 8.12 Comparison between filters manufactured on LTCC, laminate, silicon, and acoustic wave 3D glass substrate.

Fig. 8.13 The process of TGV-based bandpass filter.

deposition of the Si_3N_4 layer and the RDL process, which make up the function of the top plate of the capacitor. Finally, the RDL and passivation layer are finished to form the electrode. Fig. 8.14 shows a fabricated TGV-based bandpass filter, which consists of five 3D TGV inductors and dozens of MIM capacitors. Fig. 8.15 shows the measured results of the *S* parameters. The pass-band is centered at the frequency of 4 GHz, the insertion loss of the filter is less than 1 dB, and return loss is greater than 13 dB.

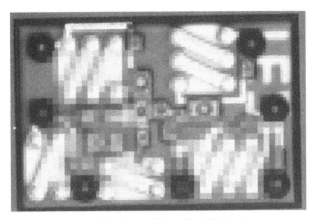

Fig. 8.14 The top-view of glass-based bandpass filter after fabrication.

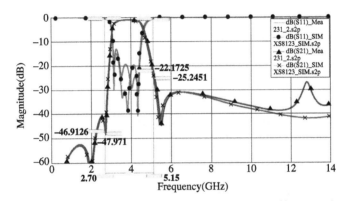

Fig. 8.15 The simulated and measured S parameters of two bandpass filter samples.

8.5 Embedded glass fan-out wafer-level package technology

8.5.1 Technology description

The fan-out wafer-level package (FOWLP) has been broadly applied in monolithic microwave integrated circuit (MMIC) packages due to the low transmission loss between chip to package [12]. Glass, as an ideal material for RF application, has been used for the carrier material of FOWLP [13]. The structure of the eGFO package is illustrated in Fig. 8.16, in which the chip is embedded in glass by lamination of the molding compound. It is a chip-first and face-down process. Because of the process flexibility of the through glass cavity, eGFO package technology can provide a solution for multichip packages.

The process flow of eGFO is illustrated in Fig. 8.17 [14]. The process begins with through glass cavities formation, which is similar to TGV formation. Then the chip is

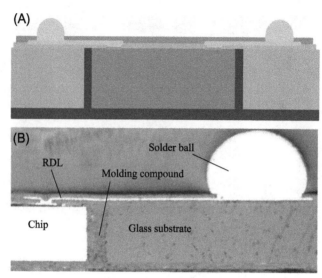

Fig. 8.16 The structure of eGFO package: (A) schematic view and (B) optical microscopy on cross section.

picked and placed into the cavity with an accuracy within 4 µm. After chip to wafer (C2W) attachment, the lamination of epoxy molding compound (EMC) is applied to fill the gap between chip and the through glass cavity. The reconstructed wafer is finished, consisting of die, glass carrier, and EMC. Then RDL is formed by thin film technology. Finally, the ball grid array and singulation are finished.

The eGFO technology has been successfully applied in a Ka-band transceiver/receiver module, which is shown in Fig. 8.18. There are three chips in the transceiver module, including a dual-channel power amplifier (PA) chip and phase shift chips in the TX module. Also, there are three chips in the RX module, including a dual-channel low-noise amplifier (LNA) chip and phase shift chips. The eGFO package technology provides a solution for multichip packages, thanks to the flexibility of the through glass cavity in the process.

8.5.2 AIP enabled by eGFO package technology

The market prospects for 5G millimeter-wave (mm-W) wireless communication, automotive radar, and passive imaging are vast. At the frequency domain of mm-W, the physical scale of the package is comparable to wavelength. Antenna in package (AiP) technology can possibly be implemented. AiP technology not only helps scale down the system and increase integration, but also reduces the transmission loss in the path to the antenna. As shown in Fig. 8.19, the antenna can be manufactured by RDL in

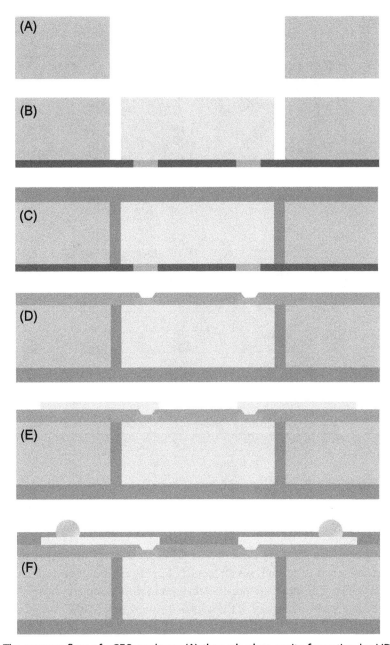

Fig. 8.17 The process flow of eGFO package: (A) through glass cavity formation by LIDE method; (B) chip to wafer placing; (C) dry film lamination; (D) debonding and first passivation; (E) patterning and copper plating; and (F) second passivation and ball formation [14].

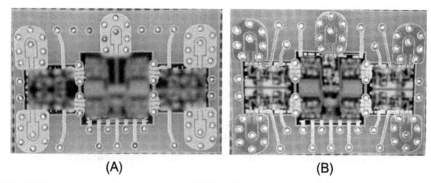

Fig. 8.18 (A) Ka-band transceiver module and (B) Ka-band receiver module.

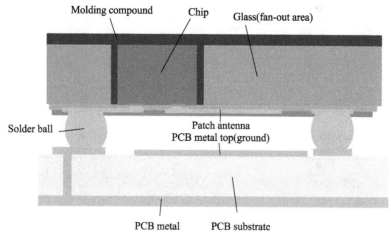

Fig. 8.19 Schematic view of 2D AiP integrated on PCB.

an eGFO package structure and the ground reflector is located on the top metal of the printed circuit board (PCB) [15].

To evaluate the electrical performance of eGFO package technology in this frequency domain, the test transmission lines with length of 1 mm are designed and measured from 73 GHz to 84 GHz. The structure parameters used in simulation are shown in Table 8.5. The designed line width and space of the coplanar waveguide (CPW) are 50 and 12 μm to realize an impedance of 50 Ω. Fig. 8.20A shows the cross section of the test transmission line and Fig. 8.20B shows the top view of CPW. Fig. 8.21 shows a simulation model and microscopic photo of CPW. The ground-signal-ground (GSG) pitch of the test probe is 150 μm. The simulation and measured results are shown in Fig. 8.22; the measured results of insertion loss are in line with the simulation results. The insertion loss measured is about 0.65 dB from 73 to 84 GHz, which is much lower than traditional package technology.

Table 8.5 Material parameter of the simulation model.

Components	Symbol	Thickness (µm)	ε_r	tan δ	μ_r	σ (S/m)
Epoxy molding compound	H_4	20	3.2–3.5 at 1 MHz	0.0067 at 1 MHz	1	0
Glass	H_3	180	5.1 at 1 GHz	0.0035 at 1 GHz	1	0
			5.1 at 5 GHz	0.0039 at 5 GHz		
			5.1 at 7 GHz	0.0049 at 7 GHz		
			5.1 at 24 GHz	0.009 at 24 GHz		
			5.0 at 77 GHz	0.011 at 77 GHz		
Dielectric	H_1, H_2	10	3.3 at 1 GHz	0.01 at 1 GHz	1	0
			3.1 at 10 GHz	0.02 at 10 GHz		
			2.9 at 25 GHz	0.02 at 25 GHz		
			2.8 at 77 GHz	0.02 at 77 GHz		
			2.8 at 100 GHz	0.02 at 100 GHz		
Redistribution layer	T	5.5	–	–	1	5.8×10^7

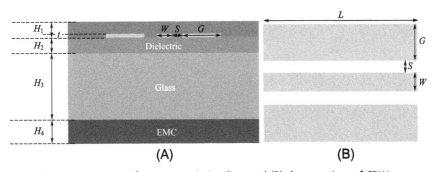

Fig. 8.20 (A) The cross section of test transmission line and (B) the top view of CPW.

Based on eGFO package technology, a 77-GHz multichannel automotive radar chip was realized. Fig. 8.23 shows the single-chip package and two-chip package with AiP. The patch antenna array is located on RDL. As illustrated in Fig. 8.24, the measured radiation patterns of the fully assembled TX module show that the total effective isotropic radiated power (EIRP) of the 3-feed antenna is improved by 7.5 dB compared with

240 TSV 3D RF integration

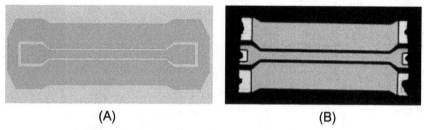

Fig. 8.21 (A) The top view of CPW simulation model and (B) microscopic photo of a fabricated CPW sample.

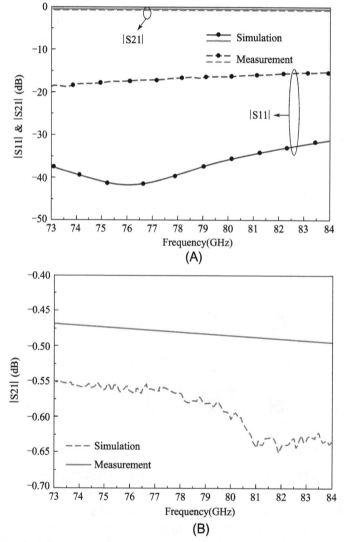

Fig. 8.22 Simulated and measured S-parameter of 1-mm CPW transmission line in the range of 73–84 GHz and (A) S11&S21 and (B) S21.

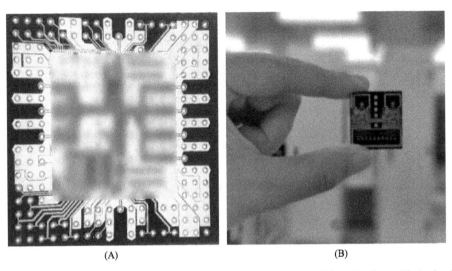

Fig. 8.23 77-GHz multichannel automotive radar chip module using eGFO technology: (A) single-chip package using eGFO package technology and (B) two-chip package with AiP [7].

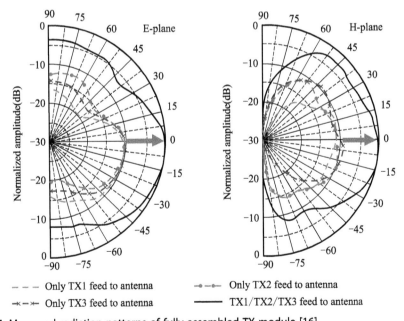

- - - - Only TX1 feed to antenna   —•—•— Only TX2 feed to antenna
—×—×— Only TX3 feed to antenna   ———— TX1/TX2/TX3 feed to antenna

Fig. 8.24 Measured radiation patterns of fully assembled TX module [16].

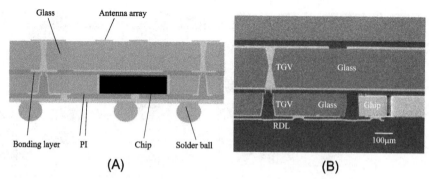

Fig. 8.25 The cointegration of the antenna in 3D eGFO package: (A) the schematic view and (B) SEM image of cross section.

the single-feed antenna. When all TXs are on and a 2 × 8 MIMO radar algorithm is optimized, the measured detection range could reach 50 m [16].

8.5.3 3D RF integration enabled by eGFO package technology

The 3D package enables high performance, miniaturized size, and high reliability of the system. The cointegration of the antenna in a 3D eGFO package is realized by TGV technology to provide a high-performance electrical connection. As is illustrated in Fig. 8.25, there are two layer glasses in which a chip is embedded and the antenna array is located on the back of the chip. The antenna is fed by a vertical TGV connection or coupling method.

8.6 2.5D heterogeneous integrated L-band receiver based on TGV interposer

Fig. 8.26 shows a schematic diagram of the 2.5D heterogeneous RF integration based on TGV interposer. Here, sandblasting is utilized to form TGV holes. Compared with methods for via formation by laser drilling or photolithography in combination with following wet etching on specialized photosensitive glass, sandblasting is a batch mode, wafer-level drilling technology with high efficiency and low-cost manufacture.

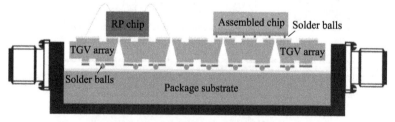

Fig. 8.26 Schematic diagram of the 2.5D RF integration-based TGV interposer.

Furthermore, sandblasting has little impact on the material properties of the substrate and it is capable of meeting the requirements of the input/output (I/O) count of RF microelectronics devices. TGV metallization is performed with conformal Cu electroplating, so has a redistribution layer on both sides at the same step, which exhibits similar electrical performance to Cu fully filled TGVs due to the skin effect and a better thermomechanical reliability behavior, while having a high filling efficiency and therefore a lower cost. The layer of Ni/Pd/Au is formed on Cu RDL selectively to address the compatibility issues of both material and process due to the fact that RF microelectronic devices are usually fabricated on III-V semiconductor materials and metallized by Au material. To evaluate the RF property of the TGV interposer presented here, a group of test vehicles was designed and fabricated. One is a 5-mm CPW transmission line where the signal line of 110 μm has a TGV electrically grounded line at both sides with a gap of 110 μm, as shown in Fig. 8.27A; the other is a CPW transmission line with similar design parameters but

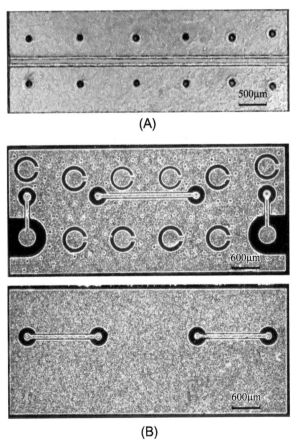

Fig. 8.27 TGV test vehicle sample: (A) CPW line grounded by TGVs and (B) CPW line linked by four RF TGVs.

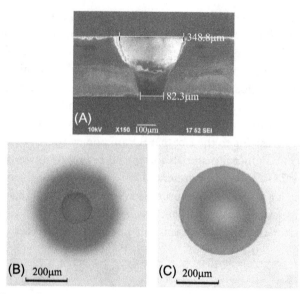

Fig. 8.28 Results of the drilled TGV after sandblasting and thinning/polishing: (A) cross-section view; (B) top-down view from exit via; and (C) top-down view from the entrance.

linked by four RF TGVs as shown in Fig. 8.27B. Fig. 8.28 shows a TGV viewed in a cross-sectional direction, bottom-up and top-down, which is formed by a sandblasting process in combination with a double-sided polishing process; a tapered profile can be found that it has an entrance diameter and exit diameter of 348.8 and 82.3 μm, respectively. Fig. 8.29 shows the RF test results. The insertion loss is 0.13 dB/mm at 10 GHz for the proposed CPW line, while the overall insertion loss is about 0.131 dB/mm at 10 GHz for the CPW line linked by four TGVs, which means less than 0.1 dB at 10 GHz for a single TGV. For both samples, the test S21 value is twice the predicted value from the HFSS model. This is ascribed to no factoring in of the surface roughness of the metal layer thickness and the thickness variation in metal layer in simulation with HFSS at the design stage. If the real surface roughness of the Cu layer is considered, a good agreement can be found in Fig. 8.29, which indicates that surface roughness should be carefully attended to for RF design or process optimization. Table 8.6 compares the tested results for CPW lines obtained from published papers, and it can be concluded that the CPW line with grounded TGVs performs better than the traditional CPW line.

In order to test the proposed TGV interposer based 2.5D heterogeneous RF integration scheme, a single-channel L-band receiver module [5], which is a simplified version of that described in section 5.2 and is composed of six chips including LNA, phase

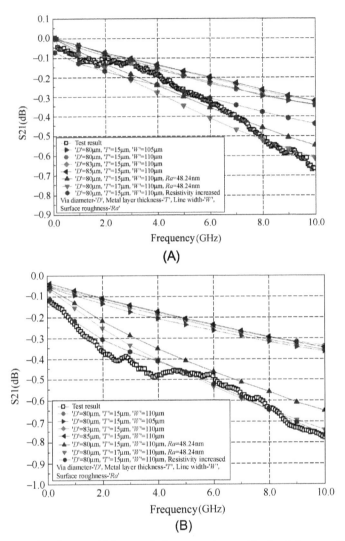

Fig. 8.29 RF test results of test vehicle sample in comparison with simulation results: (A) CPW line grounded by TGVs and (B) CPW line linked by four RF TGVs.

shifting, attenuation, and driver IC as summarized in Table 8.7, is designed, fabricated, and assembled. Fig. 8.30 shows the optical view of a fabricated TGV interposer sample for the L band receiver, while Fig. 8.31 shows the X-ray image with a cross-section view of metallized TGV with copper. Fig. 8.32 shows a photo of the 2.5D heterogeneous integrated L band receiver based on the TGV interposer, which is assembled in a similar

Table 8.6 RF test results of TGV integrated CPW lines obtained from published papers.

References	Characteristic of test structure	TGV formation method	Number of signal TGV	Metallization	S21
[8]	TGVs structure comprising two signal TGVs and four ground TGVs: 180 μm in thickness/Entrance diameter of 35 μm and exit diameter of 22 μm/50 μm in pitch (simulation)	Excimer laser	2	Cu	0.5 dB at 10 GHz (TGVs)
[9]	A CPW line to TGV transition structure: 175 μm in depth/entrance and exit diameters of 130 and 90 μm, respectively/1.6 mm in length	Excimer laser	2	Cu	<0.15 dB at 9 GHz
[12]	Microstrip models with one straight TGV: 200 μm in depth/80 μm in radius (simulation)	—	1	W	0.45 dB at 10 GHz
	RDL and two TGVs with pads: 200 μm in depth/100 μm in radius and 400 μm in pitch/the length of RDL is 1 mm (simulation)	—	2	W	−0.83 dB at 10 GHz
[13,14]	CPW transmission line: 40 μm in diameter and 80 μm in pitch/the length of 900 μm	Silicon etching, anodic bonding, glass reflow, and silicon posts removal	2	Cu	0.05 dB at 10 GHz
This work	CPW line with grounded TGVs: 110 μm in width/5 mm in length	Sand blasting	0	Cu and Au	0.13 dB/mm at 10 GHz
	CPW line with RF and grounded TGVs: entrance and exit diameters of 350 μm and 80 μm, respectively/110 μm in width/6 mm in length		4		0.131 dB/mm at 10 GHz Per TGV: <0.1 dB at 10 GHz

Table 8.7 Chips utilized for L-band receiver.

Name	LNA	Attenuator	Phase shifter	Power divider	Driver chips
Type	GNM1108	GNM6109	GNM5105	GNM7101	GN2040C
Number	1	1	1	1	2

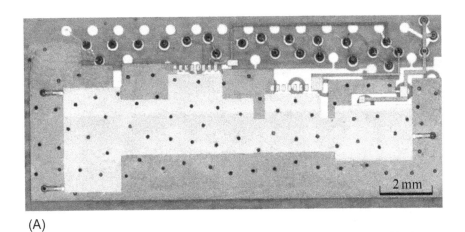

(A)

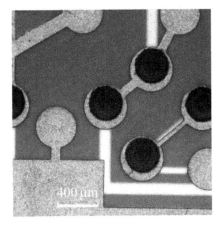

(B)

Fig. 8.30 The TGV interposer sample for L band receiver: (A) global view and (B) zoom in view of RDL on both sides.

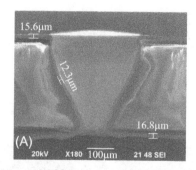

Fig. 8.31 (A) Cross section of metallized TGV by SEM; (B) overall X-ray image of the TGV interposer; and (C) detail view, respectively.

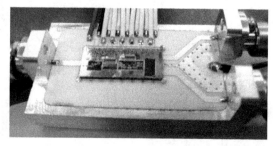

Fig. 8.32 Photo of the 2.5D heterogeneous integrated L-band receiver based on TGV interposer.

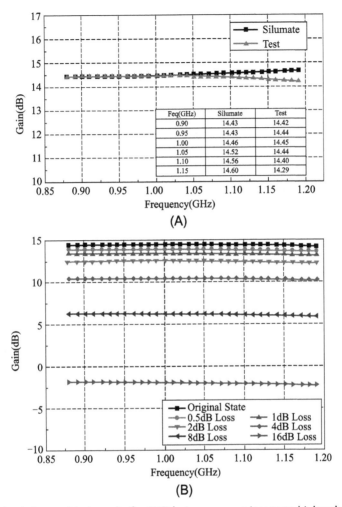

Fig. 8.33 The simulation and test results for 2.5D heterogeneous integrated L-band receiver.

way to that described in Chapter 5, section 5.2. Fig. 8.33 shows the function test results for the 2.5D heterogeneous L band receiver, the gain value is about 14.4 dB on average, which is close to the theoretical value 14.5 dB down with the extracted S parameter of the TGV sample and test S-parameters of the comprising chips.

8.7 Conclusions

This chapter discusses the experimental principles of LIDE technology and presents two methods of TGV metallization and a TGV technology-based filter. A glass embedded wafer level fan-out package is developed and implemented with a Ku-band transmitter module and a millimeter-wave AIP module. A single-channel L-band receiver based on a TGV interposer is demonstrated; the gain value is about 14.4 dB on average, which is close to the theoretical value of 14.5 dB down with the extracted S-parameter of a TGV sample and the test S-parameters of the comprising chip.

References

[1] Watanabe AO, Ali M, Sayeed SYB, et al. A review of 5G front-end systems package integration. IEEE Trans Compon Packag Manuf Technol 2020;11(1):118–33.
[2] Yole Development. Glass substrate for semiconductor applications. Yole Development; 2020.
[3] Yang Z, Wang Y, Wang H, et al. High-g MEMS shock threshold sensor integrated on a copper filling through-glass-via (TGV) substrate for surface mount application. In: 18 international conference on Slid-state sensor, Actuators and Microsystems (TRANSDUCERS), Transducers-2015, IEEE; 2015. p. 291–4.
[4] Takashashi S, Horiuchi K, Tatsukoshi K, et al. Development of Through Glass Via (TGV) formation technology using electrical discharging for 2.5/3D integrated packaging. In: Electronic components and technology conference (ECTC), 2013 proceedings 63rd; 2013. p. 348–52.
[5] Sukumaran V, Chen Q, Liu F, et al. Through-package-via formation and metallization of glass interposers. In: Electronic components and technology conference (ECTC), 2010 proceedings 60th; 2010. p. 557–63.
[6] Takahashi S, Horiuchi K, Tatsukoshi K, et al. Development of through glass via (TGV) formation technology using electrical discharging for 2.5/3D integrated packaging. In: 2013 IEEE 63rd electronic components and technology conference; 2013. p. 348–52.
[7] Ostholt R, Ambrosius N, Krüger RA. High speed through glass via manufacturing technology for interposer. In: Proceedings of the 5th electronics system-integration technology conference (ESTC); 2014. p. 1–3.
[8] Chen L, Heng W, Zhang M, et al. Development of laser-induced deep etching process for through glass via. In: 2019 IEEE international conference on electronic packaging technology (ICEPT), HongKong, China; 2019.
[9] Wei T, et al. Performance and reliability study of TGV interposer in 3D integration. In: 2014 IEEE 16th electronics packaging technology conference (EPTC); 2014. p. 601–5.
[10] Chen L, Yu T, Ren X, et al. Development of low cost through glass via (TGV) interposer with high-Q inductor and MIM capacitor. In: 2020 21st international conference on electronic packaging technology (ICEPT); 2020. p. 1–4.
[11] Qian L, Sang J, Xia Y, et al. Investigating on through glass via based RF passives for 3-D integration. IEEE J Electron Devices Soc 2018;6:755–9.
[12] Keser B, Kroehnert S. FO-WLP market and technology trends. In: Advances in embedded and fan-out wafer level packaging technologies. IEEE; 2019. p. 39–54.
[13] Ali M, et al. First demonstration of compact, ultra-thin low-pass and bandpass filters for 5G small-cell applications. IEEE Microwave Wireless Compon Lett 2018;28(12):1110–2.
[14] Yu T, Zhang X, Chen L, et al. Development of embedded glass wafer fan-out package with 2D antenna arrays for 77GHz millimeter-wave Chip. In: 2020 IEEE 70th electronic components and technology conference (ECTC); 2020. p. 31–6.
[15] Wojnowski M, Pressel K. Antenna integration in eWLB package. In: Antenna-in-package technology and applications. IEEE; 2020. p. 219–65.
[16] Duan Z, et al. 14.6 A 76-to-81 GHz 2 × 8 FMCW MIMO radar transceiver with fast chirp generation and multi-feed antenna-in-package array. In: 2021 IEEE international solid-state circuits conference (ISSCC); 2021. p. 228–30.

CHAPTER 9

Conclusion and outlook

Study results from our team on 3D heterogeneous RF integration oriented HR-Si interposer technology have been systematically covered in this book. A series of studies were conducted to solve the key problem of high-frequency loss, which included a fundamental study on designing and processing of HR-Si TSV, design and process validation of a new 3D interconnected structure based on HR-Si TSV, and application verification of 3D heterogeneous RF integration. In order to further improve the integration density, integrated inductors using HR-Si TSV and integrated patch antennas using an HR-Si interposer were explored. To overcome the heat dissipation problem caused by high-performance TR integration, a microchannel embedded HR-Si interposer was investigated. Some of the innovative results can be described as follows:

(1) CPW line with grounded TSV array workable at frequencies up to 60 GHz, and the new design of 3D interconnection for RF signal transmission were designed and developed on HR-Si interposer. The effects of key processing parameters, such as surface roughness and resistivity of copper layer on the high-frequency electrical performance, were theoretically revealed by a combination of process tests and high-frequency electrical measurement analysis. Optimized strategies were also proposed, including pseudo-coaxial TSV interconnection and redundant RF TSV interconnection, with insertion losses of 0.07 dB at 40 GHz, 0.19 dB at 40 GHz and 0.46 dB at 40 GHz, for pseudo-coaxial, dual-redundant and quad-redundant TSV interconnections, respectively, which are comparable experimental results to those of the state of the art in single TSV interconnection. Furthermore, excellent RF capability was attained with the improvement in reliability, which was proved in the 2.5D/3D heterogeneous RF integration prototypes.

(2) Developed a model of an integrated planar spiral inductor based on the HR-Si interposer as well as techniques for extracting parameters from the model. The new integrated inductor model was developed by including surface roughness as a processing parameter, which resulted in reduced modeling error and improved simulating efficiency. The main factors affecting the dimension and performance of inductors were analyzed. Based on the insurance of the potential system integration advantage and the balanced consideration of process complexity, we proposed a design of integrated local suspended planar spiral inductors on the chip to improve the quality factor of the inductor. The measured integrated inductance on the chip was 18 nH at 10 GHz, with the margin of error of 2.23 and 0.82 nH using the developed model

and HFSS full-field simulation, respectively, which achieved better precision and improved efficiency. TSV 3D inductors were designed and demonstrated. The measured inductance was 20 nH at 1.5 GHz, which effectively realized high-density inductors by improving inductance per unit area.

(3) Based on the proposed technique of HR-Si interposer, prototypes of a 2.5D integrated single-channel and quad-channel L-band receiver were developed. Microband interdigital filters with a working frequency of 7.5–10.5 GHz were realized based on the design of the HR-Si interposer. The insertion loss is about 2 dB. The follow-up 3D integrated 5–10 GHz channelized frequency conversion receivers were also developed and passed the measurement test on function and performance, which demonstrated the feasibility and advantage of the applications for HR Si interposer based 3D heterogeneous integration.

(4) With regard to the applications for 5G millimeter wave, a design of multilayer patch antennas with backside chambers based on HR-Si interposer was proposed in order to optimize bandwidth range and gain. The tests proved the feasibility of TSV interconnection based on micro-bump bonding as a feed structure. The tests on feed structures containing micro-bump, TSV and CPW revealed the insertion loss of 1.7439 dB at 40 GHz. The test on antennas gave a working frequency of 32.75 GHz, bandwidth of 1.04 GHz and a maximum gain of 3 dB. The analysis of the measurement error was also conducted.

(5) Studied the processing of a microchannel embedded HR-Si interposer based on Si-to-Si wafer level bonding. Functional and performance tests were passed on integrated microchannel cooled GaN power amplification modules.

(6) Demonstrated the capacity of TGV formation by the LIDE process, TGV metallization by the copper electroplating process, TGV-enabled inductor, filters, glass-based wafer level fanout package, and TGV interposer-based 2.5D/3D RF integration.

It is worth mentioning that the previously mentioned work was mostly a summary of our team's study on 3D heterogeneous RF integration of HR-Si/TGV interposer within the past 10 years. Although we made great efforts to review the progress of our international peers, comments and suggestions are welcome, since some missing information or mistakes made by us were unavoidable.

At present, although manufacturing of the HR-Si wafer as capping for BAW, RF and MEMS device chip-scale packages has been realized in China, this is not the case for 3D heterogeneous RF integration. There are still numerous issues remaining in the transition of HR-Si interposer to engineering applications, such as cost, throughput, yield and application reliability. We predict that this technique will enter its golden age in the next 5–10 years, even with the competition from other techniques such as through glass via interposers.

For outlook, the HR-Si interposer will advance in the following aspects: feature size scaling down, higher frequency applications, more stacking layers, more kinds of integrated chips, and integration of digital and analog signals, optical signals and cooling microfluidics. We hope more peers will contribute their efforts to Chinese chips by further developing applications of the HR-Si interposer technique for 3D heterogeneous RF integration.

Appendix 1

Abbreviations

3GPP	3rd Generation Partnership Project
5G	fifth-generation mobile communication
AC	alternating current
ADC	analog-to-digital converter
AFM	atomic force microscope
AiP	antenna-in-package
Ar	area
AR	aspect ratio
BCB	benzocyclobutene
BPF	bandpass filter
C2W	chip to wafer
CMOS	complementary metal oxide semiconductor
CMP	chemical mechanical polish
CNFET	carbon nanotube field-effect transistor
COP	coefficient of performance
COSMOS	compound semiconductor materials on silicon
CPi	capacitor parallel
CPW	coplanar waveguide
CSi	capacitor series
CTE	coefficient of thermal expansion
CVD	chemical vapor deposition
CVS	cyclic voltammetry stripping
DAHI	diverse accessible heterogeneous integration
DARPA	Defense Advanced Research Projects Agency
DC	direct current
DGS	defective ground structure
DHBT	double-heterojunction bipolar transistor
DI	deionized
DRIE	deep reactive ion etching
ECD	electrochemical deposition
EDX	energy-dispersive X-ray
eGFO	embedded glass fan-out
EIRP	effective isotropic radiated power
EM	electromagnetic field
EMC	epoxy molding compound
ENEPIG	electroless nickel electroless palladium/immersion gold
FBW	fractional bandwidth
FDM	finite difference method
FEA	finite element analysis
FOWLP	fan-out wafer level packaging
FPA	focal plane array
FPGA	field programmable gate array

GAA	gate-all-around
GSG	ground-signal-ground
HB	high band
HBT	heterojunction bipolar transistor
HEMT	high-electron-mobility transistor
HF	hydrofluoric
HFSS	high frequency structure simulator
HRP2	high reliable system profile version 2
I/O	in/out
IC	integrated circuit
ICP	inductively coupled plasma
IME	institute of microelectronics
IPD	integrated passive device
IR	infrared radiation
ISM	industrial scientific medical
K-band	Ku-band (12 – 18 GHz), u refers to under
$K_{i,i+1}$	coupling coefficient
LB	low band
L-band	low band (1 – 2 GHz)
LC	inductance/capacitance
LED	light-emitting diode
LIDE	laser-induced deep etching
LNA	low-noise amplifier
LO	local oscillator
LPF	low-pass filter
LPi	inductance parallel
LSi	inductance series
LTCC	low temperature cofired ceramic
LTE	long-term evolution
MCM	multichip module
MEMS	microelectromechanical system
MIM	metal-insulator-metal
MIMO	multiple input multiple output
MMIC	monolithic microwave integrated circuit
MMPA	Minnesota Municipal Power Agency
MW	microwave
N	number of snowballs
NEMS	nano-electromechanical system
NRL	Naval Research Laboratory
PA	power amplifier
PCB	printed circuit board
PI	photosensitive polyimide photoresist
PMR	private mobile radio
PN	PN junction
PRAM	programming in unipolar resistive-switching memory
PVD	physical vapor deposition
Q	quality factor

Q-COV	quasi-coaxial-via
RCA	Radio Corporation of America
RDL	redistribution layer
RF	radio frequency
RF-SIP	radio frequency-system in package
RIE	reactive ion etching
RMS	root-mean-square
S11	reflection coefficient
S21	insertion loss
SAP	semiadditive process
SEM	scanning electron microscope
SIP	system-in-package
SIW	substrate integrated waveguide
SMD	surface mount device
SOC	system on chip
SOI	silicon on insulator
SOLES	silicon on lattice engineered substrate
SOLT	short open load through
S-parameter	scatter parameter
SU	ultraviolet process
TGV	through glass via
TM	transverse magnetic
TR	transmitter/receiver
TSC	through silicon capacitor
TSV	through silicon via
Tx/Rx	transmitter and receiver
UBM	under bump metallization
UV	ultraviolet
UWB	ultrawide band
VCO	voltage-controlled oscillator
VGA	video graphics array
VMS	virgin makeup solution
VNA	vector network analyzer
VSWR	voltage standing wave ratios

Appendix 2

Nomenclature

Chapter 2: Design, process, and electrical verification of high-resistivity Si interposer for 3D heterogeneous RF integration

Nomenclature (units of measure)

$\vec{J}$	current density (A/m²)
a	the outer radius of the Cu tube (μm)
A_r	electrical charge (C)
b	the outer radius of the Si substrate (μm)
c	the inner radius of the Cu tube (μm)
E	elastic modulus (MPa)
Freq.	frequency (Hz)
I	current (A)
k	the inner and outer radius ratio of the Cu tube
m	mass (g)
M	molar mass
Q	electrical charge (C)
q	the ratio of the outer radius of silicon to the outer radius of the copper pillar
R	radius (μm)
r	thermal stress location (μm)
S	area (dm²)
S	scattering parameters (dB)
S11, S21	reflecting parameter and insertion loss
T	temperature (K or °C)
T	thickness (μm)
t	time (s)
ΔT	uniform temperature load (°C)

Greek letters

α	thermal expansion coefficient
δ	stress (MPa)
ν	Poisson's ratio
τ	shear stress (MPa)

Chapter 3: Design, verification, and optimization of novel 3D RF TSV based on high-resistivity Si interposer

Nomenclature (units of measure)

h_{tooth}	peak height (μm)
A_{lat}	the surface area of the protrusion (μm)
A_{sphere}	the surface area of the individual sphere (μm)
A_{title}	the area of the surrounding plane (μm)
K_{Huray}	the surface roughness correction factor

b_{base} peak bottom width (μm)
d_{peaks} the distance between the peaks (μm)
c speed of light (km/s)
f frequency (Hz)
k wave number
Re   Reynolds number

Greek letters

ε_0 vacuum dielectric constant (F/m)
μ_0 vacuum permeability (H/m)
δ skip depth (μm)
ε_{Si} relative dielectric constant of silicon
η intrinsic impedance (Ω)
λ wavelength (m)
μ permeability (H/m)
ω angular frequency (rad/s)

Chapter 4: HR-Si TSV integrated inductor

Nomenclature (units of measure)

AMD   algebraic average distance (m)
B   magnetic strength (T)
C_{air} coupling capacitance introduced by the cavity structure (F)
C_{OX} insulation layer capacitance (F)
C_S feed-through capacitance (F), capacitance of parasitic capacitor (F)
C_{Si} coupling capacitance of Si substrate (F)
C_{Sub} substrate capacitance (F), capacitance of parasitic capacitor (F)
C_{TSV} feed-through capacitance between TSV and the planar spiral structure (F)
d distance between center lines (m)
d distance between the two TSVs (m), the length of common perpendicular to the two RDLs (m)
$d_{T,diff}$ distance between the two TSVs at different rings (m)
$d_{T,in}$ distance between the two TSVs at inner rings (m)
$d_{T,out}$ distance between the two TSVs at outer rings (m)
GMD   geometric mean distance (m)
GMD_M geometric mean distance between i and j (m)
G_{Sub} substrate conductance (S), conductance of parasitic conductance (S)
h height of through silicon via (μm)
I_{eddy} eddy current (A)
I_{ex} excitation current (A)
J imaginary unit
K_{Huray} surface roughness factor calculated by using Huray model
$K_{Huraymax}$ maximum surface roughness factor using classic Huray model
l total length of the center lines (m)
L   inductance (H)
l_i length of i-th segment of straight wire (m)
L_i self-inductance of i-th segment (H), length of the i-th segment (m)
$L_{i,tot}$ total self-inductance of RDLs (H)
L_j length of the j-th segment (m)

l_{Rb}	length of bottom RDL (m)
l_{Rt}	length of top RDL (m)
L_S	self-inductance (H)
l_T	length of TSVs (m)
$M(i,j)$	mutual inductance between ith segment of straight wire and jth segment of straight wire (H)
$M_{R,np}$	mutual inductance for nonparallel RDLs (H)
$M_{R,p}$	mutual inductance for parallel RDLs (H)
M_T	partial mutual inductance of two TSVs (H)
N	total count
p	distances between the ends of j (m)
P_{flat}	flat profile
P_n	loss power (W)
$P_{spheres}$	spherical profile
q	distances between the ends of i (m)
Q	quality factor
R	resistance (Ω)
R_{eddyn}	eddy current impedance of nth metal segment (Ω)
R_i	radius of the inner ring (m)
R_n	DC impedance of n-th metal segment (Ω)
R_o	radius of the outer ring (m)
R_S	self-impedance (Ω)
R_{sheet}	unit resistance after splitting (Ω)
R_{Si}	leakage resistance corresponding to C_{Si} (Ω)
r_T	radius of TSVs (m)
S	conductor cross-sectional area (m^2)
s	endpoints offset of the two RDLs (m)
T	frequency correction factor
t	thickness of metal spiral coil (m)
t_{ox}	thickness of the insulator material (m)
t_R	thickness of RDL (m)
U_X	mutual inductance parameter
w	line width of metal spiral coil (m)
w_R	width of RDL (m)
Y_S	conductance of spiral inductor
Y_{Sub}	conductance of substrate

Greek letters

α	extension lengths to the intersection point P(m)
β	extension lengths to the intersection point P(m)
γ	a coefficient associated with the proximity effect after surface bending
δ	skin depth at high frequency (m)
ε	distributed angle around the central axis (rad)
ε_0	vacuum dielectric constant (F/m)
ε_{ox}	dielectric constant of the insulator material
η	angle of top RDL and adjacent bottom RDL (rad)
Θ	inclination angle (rad)
μ	magnetic permeability (H/m)
μ_0	vacuum permeability (H/m)
ρ	resistivity of the metal ($\Omega\cdot$m)

σ	electrical conductivity (S/m)
τ	factor that varies with frequency
φ	rotation angle (rad)
ω	angular frequency (rad/s)
Ω	solid angle (sr)
ω_{Qmax}	frequency when quality factor is maximum (Hz)
ϖ	angular frequency (rad/s)

Chapter 5: Verification of 2.5D/3D heterogeneous RF integration of HR-Si interposer

Nomenclature (units of measure)

f_0	resonant frequency (Hz)
g_i	normalized elements
h	height(mm)
$K_{i,i+1}$	coupling coefficient
L	length (m)
R_L	resistance of load (Ω)
R_S	resistance of source (Ω)
V	voltage (V)
W	width (mm)
Y	admittance (Ω^{-1})

Greek letters

e	effective dielectric constant

Chapter 6: HR-Si interposer embedded microchannel

Nomenclature (units of measure)

$\dot{m}$	mass flow rate in the microchannel (kg)
$2a$	heat source diameter (m)/uniform heat source length (m)
$2b$	uniform heat source width (m)
$2c$	uniform substrate length (m)
$2d$	uniform substrate width (m)
A	cross-sectional area of heat source (m^2)
a_1	width of cross-section of one microchannel (m)
A_{bump}	cross-sectional area of bump (m^2)
A_{TSV}	cross-sectional area of TSV (m^2)
A_{wire}	cross-sectional area of wire (m^2)
b_1	height of cross-section of one microchannel (m)
C	heat capacity of the cooling liquid (J/(kg·K))
c_1	distance of adjacent microchannels (m)
C_p	specific heat capacity of DI water (J/(kg·K))
d	length of the flow channel (m)
D_h	the feature size (m)
h	convective heat transfer coefficient (W/(m^2·K))
h_{equ}	equivalent heat transfer coefficient (W/(m^2·K))
I_n	input current (A)

k	thermal conductivity (W/(m·K))
k_1	thermal conductivity of the chip1 (W/(m·K))
k_2	thermal conductivity of the Si interposer (W/(m·K))
k_{bump}	thermal conductivity of bump (W/(m·K))
k_f	thermal conductivity of the fluid (W/(m·K))
k_s	substrate thermal conductivity (W/(m·K))
k_{Si}	thermal conductivity of Silicon (W/(m·K))
k_{sub}	thermal conductivity of the substrate (W/(m·K))
k_{TSV}	thermal conductivity of TSV (W/(m·K))
$k_{underfill}$	thermal conductivity of underfill (W/(m·K))
k_{wire}	thermal conductivity of wire (W/(m·K))
Nu_{ave}	average Nusselt number
P	pressure (Pa)/ pump power (W)
P_n	maximum heat dissipation power (W)
Q_{max}	power dissipation of the system (W)
Q_n	maximum dissipation power (W)
R	thermal resistance (K/W)
R_0	initial resistance value of customized heat source chip (Ω)
R_{1D-sub}	one-dimensional thermal resistance of the substrate (K/W)
R_{base}	heat transfer resistance of the substrate (K/W)
$R_{base,conv}$	heat transfer resistance due to the substrate convection (K/W)
r_{bump}	radius of bump (m)
$R_{bump-1D}$	thermal resistance of the differential section of bump (K/W)
$R_{bump-sp}$	diffusion thermal resistance of bump (K/W)
R_{chip}	one-dimensional conductive thermal resistance of the chip (K/W)
R_{fluid}	thermal resistance of the coolant (K/W)
$R_{microchannel}$	thermal resistance of the microchannel (K/W)
R_{model}	total thermal resistance of the chip1 (K/W)
r_{Si}	radius of Si (m)
R_{sp}	diffusion thermal resistance (K/W)
R_T	recorded resistance value during the test (Ω)
R_{tot}	total thermal resistance between the junction temperature of the device and the coolant inlet temperature (K/W)
r_{TSV}	radius of TSV (m)
R_{TSV-1D}	thermal resistance of the differential section of TSV (K/W)
R_{TSV-sp}	diffusion thermal resistance of bump (K/W)
R_{wall}	heat transfer resistance of the inside wall of microchannel (K/W)
$R_{wall,conv}$	heat transfer resistance of the inside wall of microchannel facing (K/W)
$R_{wire-1D}$	thermal resistance of the differential section of wire (K/W)
s	cross-sectional area of the microchannel (m^2)
T	temperature (K)
t	thickness of the chip (m)
t_1	thicknesses of the first layer (m)
t_2	thicknesses of the second layer (m)
t_3	thickness of Si film above microchannel (m)
T_a	temperature of coolant at inlet region (K)
T_i	inlet temperature (K)
T_j	temperature rise of chip (K)

T_o outlet temperature (K)
T_{or} temperature coefficient of the customized heat source chip (K^{-1})
T_{ref} reference temperature (K)
V flow rate (mL/min)
v working fluid velocity (m/s)
V_n input voltage (V)
w thickness of dielectric layer 1 (m)

Greek letters
α aspect ratio
ρ fluid density (kg/m^3)

Chapter 7: Patch antenna in stacked HR-Si interposers
Nomenclature (units of measure)
L length of the patch (m)
W width of the patch (m)
H thickness of the substrate (m)
b length of the general patch antenna (m)
c speed of light (m/s)
f_0 operating frequency of the antenna (Hz)
l' position of the feed point of the antenna along the length of the patch (m)
S scatter parameter

Greek letters
λ wavelength in the medium (m)
ε_r relative permittivity of the antenna's dielectric
λ_e wavelength of the guided wave in the medium (m)
δ_1 length of the antenna equivalent radiation slot (m)
ε_e effective dielectric constant (F/m)
θ direction angle (degrees)

Chapter 8: Through glass via technology
Nomenclature (units of measure)
D TGV diameter (μm)
E laser energy (μJ)
f frequency (Hz)
H TGV height (μm)/substrate thickness (μm)
h_{bottom} bottom plate thickness (μm)
h_{top} top plate thickness (μm)
L inductance (H)
N number of turns
P loop pitch (μm)
S overlap area of capacitor (μm^2)
T TGV thickness (μm)/dielectric thickness (μm)
W RDL width (μm)
Y_{11} Y parameter

Greek letters

τ_p pulse width (ps)
ε_r relative dielectric constant
ε_{r1} relative dielectric constant of Si_3N_4
δ loss angle
μ_r relative permeability
σ conductivity (S/m)

Appendix 3

Conversion factors

1 meter (m)	= 3.3 feet (ft)
1 foot (ft)	= 12 inches (in)
1 inch (in)	= 2.54 centimeters (cm)
1 ASD	= 1 A/dm^2
1 ASF	= 1 A/ft^2
1 ASD	= 9.29 ASF
1 gallon (gal)	= 16 cups
1 cup	= 237 milliliters (mL)
1 kilogram (kg)	= 2.2 pounds (lbs)
1 atmosphere (ATM)	= 101.3 kilopascals (kPa)
	= 760 millimeters of mercury (mmHg)

Index

Note: Page numbers followed by *f* indicate figures and *t* indicate tables.

A
Agilent vector network analyzer, 213–217
Air tightness test, 189
Analog signals, 24
Annealing treatment, 89–90

C
Carbon nanotube field-effect transistors (CNFET), 3
Cascade Microtech Infinity probe, 217
Cascade Microtech Summit, 217
Channelized frequency conversion receiver, 147*f*
Chebyshev T-type normalized equivalent low-pass circuit, 135*f*
China Electronics Technology Group Corporation (CETC), 153, 198
Coaxial cable, 200
Coaxial-like redundant TSV structure, 143–145
Coaxial-like transmission structure, 69, 69–70*f*
Coefficient of performance (COP), 177–178
Coefficient of thermal expansion (CTE), 86–88
Cold laser category, 17
Convex-concave structure, 155
Cooling capacity characterization, 174–176*f*, 176–178
Cooling microchannel, 178–188, 178–182*f*, 184–186*f*
Cooling microfluidics, 24
Coplanar waveguide (CPW), 38, 38*f*, 41*f*, 86, 86*f*, 87*t*, 88–89, 143*f*, 251
Copper conductive paste, 21
Copper layer, 74–76
CPW. *See* Coplanar waveguide (CPW)
Cu electroplating, 45–53, 46*f*, 48*f*

D
2.5D/3D integrated receiver, HR-Si interposer
  channelized frequency conversion receiver, 147*f*
  Chebyshev T-type normalized equivalent low-pass circuit, 135*f*
  coaxial-like redundant TSV structure, 143–145
  CPW simulation isolation, 143*f*
  design, 142–145
  electroplated Cu layer, 140–142
  fabrication, 142–145
  filter size diagram, 136*f*
  four-channel 2.5D heterogeneous integrated L-band receiver, 126–131, 127*f*, 129–130*f*, 132*t*
  high isolation, 142–143
  HR-Si interposer integrated microstrip interdigital filter, 134–142, 134–135*f*
  interdigital filter, 134*f*
    equivalent circuit model of, 140*f*
  isolation analysis, 143–145
  low temperature cofired ceramics (LTCC), 125–126
  material parameters, 136
  microstrip interdigital filter, 134
  multichip modules (MCMs), 125–126
  physical image, 138*f*
  sample electromechanical performance test, 148*f*
  test, 142–145
  3D heterogeneous integrated assembly and test, 145–150, 147–148*f*
  3D heterogeneous integrated channelized frequency conversion receiver, 130–132*f*, 132–150, 133*t*
Deep reactive ion etching (DRIE), 27, 43–44, 44*f*, 172
Defective ground structure (DGS), 198
Device chip-scale packages, 252
  3D heterogeneous integrated assembly and test, 145–150, 147–148*f*
  3D heterogeneous integrated channelized frequency conversion receiver, 130–132*f*, 132–150, 133*t*
Dielectric constant value, 79
Digital signals, 24
Dimensional parameters, 71*t*
Direct electroless copper plating, 20–21
Double-sided deep reactive ion etching (DRIE), 43–44, 44*f*

E

Eddy current loss, 102, 103f
Electroless nickel electroless palladium immersion gold (ENEPIG), 54, 54f
Electromagnetic (EM) field energy, 98–99, 98f
Electromagnetic waves, 200
Electronic Components and Technology Conference (ECTC), 8–9
Electroplated Cu layer, 140–142
Embedded glass fan-out (eGFO) package technology
  AIP enabled by, 236–242, 238f, 239t, 240–241f
  technology description, 235–236, 236–237f
Equivalent circuit model, 99f
Equivalent network model, 106f
Equivalent thermal resistance network, 164–171
External fluid, 154
External grounding conductor, 65

F

Fabrication process, 31–38, 225–228, 226t, 226f, 228f, 228t
Finite difference method (FDM), 76
5D heterogeneous integrated L-band receiver, 242–249, 242–245f, 246–247t
Form firm insulation layer, thermal oxidation to, 44–45, 45t
Four-channel 2.5D heterogeneous integrated L-band receiver, 126–131, 127f, 129–130f, 132t
Functional verification, 189–190

G

GaN device design, 158
GaN layer, 156
Gate-all around (GAA) transistors, 1
Georgia Institute of Technology, United States, 21–23
Greenhouse method, 99
Groisse model, 76–77, 77f

H

Heat dissipation capacity, 177
Heterojunction bipolar transistor (HBT), 3
HFSS software, 107, 112–113
High-frequency (HF) electrical characteristics, 28–29
High-frequency measurement, 81
High-resistivity silicon (HR-Si) interposers, 198
High-resistivity Si (HR-Si) substrate process
  coplanar waveguide (CPW), 38, 38f, 41f
  Cu electroplating, 45–53, 46f, 48f
  deep reactive ion etching (DRIE), 27
  design, 31–38, 32f
  double-sided deep reactive ion etching (DRIE), 43–44, 44f
  electroless nickel electroless palladium immersion gold (ENEPIG), 54, 54f
  fabrication process, 31–38
  form firm insulation layer, thermal oxidation to, 44–45, 45t
  high-frequency (HF) electrical characteristics, 28–29
  HR silicon, 27
  material parameters, 32t
  RF transmission structure, design and analysis, 38–43, 38f
  Silex, 27–28
  surface passivation, 54–55, 55f
  technical parameters for, 37t
  transmission structure, electrical characteristics analysis of, 55–61, 56t, 56–57f, 59t, 59f, 62t
High-resistivity Si (HR-Si) TSV interposer
  annealing treatment, 89–90
  coaxial-like transmission structure, 69, 69–70f
  coefficient of thermal expansion (CTE), 86–88
  copper layer, 74–76
  CPW, 86, 86f, 87f, 88–89
  dielectric constant value, 79
  dimensional parameters, 71t
  external grounding conductor, 65
  finite difference method (FDM), 76
  Groisse model, 76–77, 77f
  high-frequency measurement, 81
  Huray snowball model, 77
  ICT Device & Packaging Research Center, 65–68
  insertion loss precise value, 80–81
  internal transmission conductor, 65
  metal surface roughness, 88t
  optimization, 83–90, 84f
  Q-COV coaxial, 65–68
  quad-redundant TSV fault simulation results, 72f
  redundant RF TSV transmission structure, 70–72, 71f
  resistivity measurement, 74t
  root-mean-square (RMS), 76
  sample processing, 72–83

simulation error, 80
S parameters, 70–72, 71f, 86, 86f
surface treatment, 85–86, 85f
test result analysis, 72–83
TSV de-embedding method, 81–83
High temperature co-fired ceramic (HTCC), 2
HR-Si interposer embedded microchannel, 253
  application verification of, 188–190, 190–191f
  China Electronics Technology Group Corporation, 153
  convex-concave structure, 155
  cooling capacity characterization, 174–176f, 176–178
  cooling microchannel, 178–188, 178–182f, 184–186f
  design, 158–160, 159–160f
  external fluid, 154
  GaN layer, 156
  LTCCs, 155–156
  microchannels, 154–155
  microfluidics, 154
  monolithic microwave integrated circuit (MMIC), 157
  Nusselt number, 155
  process development, 172–175, 172–173f
  Reynolds number, 155
  SiC-based microchannel heat sink, 156–157
  solder-bonded diamond substrate, 157
  thermal characteristics analysis, 161–171, 161f
    analytical formula, direct calculation based on, 163–164, 163f
    equivalent thermal resistance network, 164–171, 165f, 167f, 170t
    variable diffusion angle, simplified calculation based on a, 162–163, 163f
  thermal stress issues, 157
  transmitter/receiver components, 153
HR-Si interposer integrated microstrip interdigital filter, 134–142, 134–135f
HR-Si interposer technology, 7–15, 8–11f, 12t, 13f, 15f
  carbon nanotube field-effect transistors (CNFET), 3
  gate-all around (GAA) transistors, 1
  high-resistivity Si (HR-Si) interposer, 2
  high temperature co-fired ceramic (HTCC), 2
  low temperature co-fired ceramic (LTCC), 2
  radio frequency (RF), 1

TGV interposer technology, 16–23, 18–21f, 22t, 23f
3D RF heterogeneous integration scheme, 2–6, 4f
through glass via (TGV) interposer technologies, 2
through silicon via (TSV) 3D integration technology, 1
transistor level, 5–6
ultrawide band (UWB), 3–5
HR-Si TSV integrated inductor
  eddy current loss, 102, 103f
  electromagnetic (EM) field energy, 98–99, 98f
  equivalent circuit model, 99f
  equivalent network model, 106f
  Greenhouse method, 99
  HFSS software, 107, 112–113
  Huray model, 104
  low-frequency range, 104–105
  Matlab simulation parameter value, 108t
  Matlab software, 107
  metal and nonmetal surfaces, 103
  metal wires flows, 101
  planar inductor, 96–113, 97t
  planar spiral inductor, 96–98, 98f
  planar spiral inductors, 107, 107f
  process-related parameters, 107
  proximity effect, 112–113
  real inductor, 102
  self-inductance of, 99
  silicon-based integrated inductors, 95
  3D inductor based on, 113–122, 114f, 116f, 118–120f
Huray model, 77, 104

I

ICT Device & Packaging Research Center, 65–68
Insertion loss precise value, 80–81
Integrated circuit (IC) industry, 1
Interdigital filter, 134f
  equivalent circuit model of, 140f
Internal transmission conductor, 65
Isolation analysis, 143–145

L

Laser ablation, 17
Laser-induced etching method, 17–18
Long-term evolution (LTE) system, 11
Low-cost filling solution, 20–21

Low-frequency range, 104–105
Low-loss tangent materials, 11–14
Low temperature cofired ceramics (LTCC), 2, 125–126

M
Material parameters, 32t, 136
Matlab simulation parameter value, 108t
Matlab software, 107
Metal conductive paste, 21
Metallization, 228–229, 229–230f
   metal/nonmetal surfaces, 103
   metal surface roughness, 88t
   metal wires flows, 101
Michigan State University, 198
Microstrip
   antennas, 200
   interdigital filter, 134
   line, 200
Millimeter-wave antenna, 209
MIM capacitor, 230–231, 231t, 231–232f
Monolithic microwave integrated circuit (MMIC), 2–3, 157
Multichip modules (MCMs), 125–126
Multifrequency multimode power amplifiers, 11
Muzychka's heat sink, 168

N
Nusselt number, 155

O
Optical signals, 24

P
Parallel microchannel, 166
Parametric simulation analysis, 207–209
Passive devices
   MIM capacitor, 230–231, 231t, 231–232f
   technology description, 230
   TGV-based bandpass filter, 231–234, 234f
PCB board-level integration methods, 197–198
Planar inductor, 96–113, 97t
Planar spiral inductors, 96–98, 98f, 107, 107f
Plasma etching method, 17
Polymerization reaction, 172–173
Printed circuit board (PCB), 9–11
Process-related parameters, 107
Proximity effect, 112–113
Pseudo-coaxial TSV interconnection, 251

Q
Q-COV coaxial, 65–68
Quad-redundant TSV fault simulation results, 72f

R
Real inductor, 102
Redistribution layer (RDL), 18–20
Redundant RF TSV transmission structure, 70–72, 71f
Reliability test, 189
Research Institute of China Electronics Technology Group Corporation, 14–15
Resistivity measurement, 74t
Reynolds number, 155
RF transmission structure, design and analysis, 38–43, 38f
Root-mean-square (RMS), 76

S
Sample electromechanical performance test, 148f
Sample processing, 72–83
SiC-based microchannel heat sink, 156–157
Silex, 27–28
Silicon-based integrated inductors, 95
Silicon materials, 198
Silicon-silicon bonding, 174f
Simulation error, 80
Solder-bonded diamond substrate, 157
S parameters, 70–72, 71f, 86, 86f
Stacked high-resistivity Si interposers, patch antenna in
   China Electronics Technology Group Corporation (CETC), 198
   defective ground structure (DGS), 198
   design, 200–213, 202–203t
   millimeter-wave antenna, 209
   parametric simulation analysis, 207–209
   processing, 213
   silicon materials, 198
   test and analysis, 213–222, 214f, 218f
   theoretical basis, 200, 201f
Standard RCA cleaning, 173
Surface passivation, 54–55, 55f
Surface treatment, 85–86, 85f

T
Temperature shock test, 189
Temperature storage experiment, 189

Thermal characteristics analysis, embedded
    microchannel, 161–171, 161f
  analytical formula, direct calculation based on,
      163–164, 163f
  equivalent thermal resistance network, 164–171,
      165f, 167f, 170t
  variable diffusion angle, simplified calculation
      based on a, 162–163, 163f
Thermal stress issues, 157
Through glass via (TGV) technology
  embedded glass fan-out (eGFO) package
      technology
    AIP enabled by, 236–242, 238f, 239t, 240–241f
    technology description, 235–236, 236–237f
    3D RF integration enabled by, 242, 242f
  fabrication, 225–228, 226t, 226f, 228f, 228t
  5D heterogeneous integrated L-band receiver,
      242–249, 242–245f, 246–247t
  metallization, 18–20, 228–229, 229–230f

passive devices
  bandpass filter, 231–234, 234f
  MIM capacitor, 230–231, 231t, 231–232f
  technology description, 230
Through silica via (TSV) technology
  coaxial-like redundant structure, 143–145
  de-embedding method, 81–83
  high-resistivity Si interposer (*see* High-resistivity
      Si TSV interposer)
  3D integration technology, 1
Toroidal inductor, 119
Transmission structure, electrical characteristics
    analysis of, 55–61, 56t, 56–57f, 59t, 59f, 62t
Transmitter/receiver components, 153
Two-TSC matrix capacitor, 9–11

U
Ultrawide band (UWB), 3–5